D0908663

Alan Stern and Jacqueline Mitton
Pluto and Charon

Alan Stern and Jacqueline Mitton
Pluto and Charon

*Ice Worlds on the Ragged Edge
of the Solar System*

Second Edition, Revised and Updated

WILEY-VCH Verlag GmbH & Co. KGaA

The authors

Alan Stern
Boulder, Colorado
USA

Jacqueline Mitton
Cambridge
UK

All books published by Wiley-VCH are carefully produced. Nevertheless, authors, editors and publisher do not warrant the information contained in these books, including this book, to be free of errors. Readers are advised to keep in mind that statements, data, illustrations, procedural details or other items may inadvertently be inaccurate.

Library of Congress Card No.:
applied for

British Library Cataloguing-in-Publication Data:
A catalogue record for this book is available from the British Library.

**Bibliographic information published by
Die Deutsche Bibliothek**
Die Deutsche Bibliothek lists this publication in the Deutsche Nationalbibliografie; detailed bibliographic data is available in the Internet at http://dnb.ddb.de

© 2005 WILEY-VCH Verlag GmbH & Co KGaA, Weinheim

Typesetting: Steingraeber Satztechnik GmbH, Dossenheim
Printing and Binding: Ebner & Spiegel GmbH, Ulm
Cover Design: Himmelfarb, Eppelheim, www.himmelfarb.de

Printed in the Federal Republic of Germany
Printed on acid-free paper

ISBN-13: 978-3-527-40556-5
ISBN-10: 3-527-40556-9

Contents

Pluto and Charon, S. Alan Stern and Jacqueline Mitton
Copyright © 2005 WILEY-VCH Verlag GmbH & Co. KGaA, Weinheim
ISBN: 3-527-40556-9

Preface to the Second Edition

With the proposed – and now imminent – 2006 launch of NASA's *New Horizons* Pluto–Kuiper Belt mission, we have updated and expanded *Pluto and Charon*. Since the first edition, which was mostly written in 1996 and so is now almost a decade old, many exciting developments have taken place in the study of both Pluto–Charon and the Kuiper Belt. These range from the discovery of dramatic changes in Pluto's atmosphere, to the discovery of exotic ices on Pluto's moon Charon, to the cataloguing of over 1000 Kuiper Belt Objects – some almost half the size of Pluto and partnered by satellites of their own. We have brought our book up to date by chronicling these and other recent advances in the study of Pluto–Charon and the Kuiper Belt. Additionally, we have written a new chapter, detailing the story of how *New Horizons* came into existence, its budgetary trials, its capabilities for exploring Pluto–Charon and the Kuiper Belt, and its planned mission. We eagerly anticipate a third edition, another decade hence, when the results from *New Horizons* will have revolutionized our view of Pluto, the "ninth sister" among the known planets.

18 February 2005   *S. Alan Stern and*
75th anniversary   *Jacqueline Mitton*
of the discovery of Pluto

Postscript

As this revised edition of our book was in press, we happily learned of the discovery of a remarkable body, provisionally called 2003 UB313 until it receives its formal name. "UB313" orbits beyond Pluto and is a bit larger than Pluto itself. On average, its distance from the Sun is about three times Pluto's. This body confirms predictions, which we describe in Chapter 6, that many more planets, most of them ice dwarfs, would be found in the deep outer solar system. We

Pluto and Charon, S. Alan Stern and Jacqueline Mitton
Copyright © 2005 WILEY-VCH Verlag GmbH & Co. KGaA, Weinheim
ISBN: 3-527-40556-9

suspect that, in the coming decade, many more ice dwarf planets will be discovered, along with planets of much greater size. A revolution is indeed upon us as we come to realize that our solar system has many, many more planets than the nine we have known about since our youth. Once again, Pluto has proved to be a pivotal body for understanding the content and origin of our solar system.

1 August 2005   *S. Alan Stern and*
                                           Jacqueline Mitton

Preface to the First Edition

This book was born in the summer of 1994, when Jacqueline Mitton interviewed Alan Stern about Pluto for a BBC radio program. The two of us immediately hit it off, and began exploring how we might pursue a proposal for a book about Pluto. Alan had been contemplating an idea for such a book for some years. With Jacqueline's experience of writing and editing astronomy books and Alan's intimate knowledge of Pluto, we thought we'd have a good chance of putting together something that would be entertaining, engaging, and fully up-to-date. After all – no book on Pluto had been written since 1980 – the veritable dark ages of Pluto studies.

We developed the book proposal together, and it was accepted by John Wiley & Sons in late 1995. We then divided the work. Alan did the writing; Jacqueline provided critiques of the text as the chapters were completed, and took on the enormous job of creating and gathering illustrations. The text was drafted and the illustrations collected during the first half of 1996. The material was reviewed and revised during the remainder of 1996. To ensure accuracy and balanced treatment, we asked various scientific experts and lay people to read different chapters or pairs of chapters. Every chapter benefited from four such critiques. Additionally, the whole text was read by John Wiley & Sons' technical referees, Marc Buie and Steve Shore, who were both tremendously helpful. We are indebted to all of these readers.

One thing that was difficult for us was figuring out a method of handling the Pluto research in which Alan himself had been involved. Because there are two of us, it wasn't right to use the first person, "I," nor was it correct to say "we." It also didn't feel right to us or to our publisher to refer to this work as if it were someone else's – that is, as "Stern's work." So what did we do? We simply referred to the work without direct attribution. If you are curious about which Pluto research discussed in this book Alan was involved in, the primary ones were the question and resolution of why Pluto is bright, the

Pluto and Charon, S. Alan Stern and Jacqueline Mitton
Copyright © 2005 WILEY-VCH Verlag GmbH & Co. KGaA, Weinheim
ISBN: 3-527-40556-9

measurement of Pluto's brightness temperature at IRAM in 1993, the search for small satellites around Pluto, the exploration of Pluto's dynamics with Hal Levison, the "Thousand Plutos" concept, some of the recent modeling work on the origin and evolution of the Kuiper Belt, and the Hubble Space Telescope (HST) imaging and mapping project for Pluto.

Our goal has been to write an up-to-date, but rather different kind of a book about Pluto. We wanted to create something that would be useful to students of astronomy and planetary science, but would also be interesting for the layperson. In fact, it is our hope that the majority of our readers will be astronomy enthusiasts who see Pluto as we do – as a paradigm for the modern approach to exploring the planets.

We wanted to thread three themes through the writing. The first is the story of the advances made possible by the recent technological explosion in ground-based astronomical instrumentation. Since the late 1970s, new instruments (CCD imagers, infrared spectrographs, and spacecraft observatories) have dramatically increased the grasp of the telescopes with which the planets can be explored from Earth and Earth-orbit by a factor of a thousand. The impact of such devices on planetary astronomy is best illustrated by their use on Pluto, precisely because Pluto only became a subject of serious study as a result of this revolution in astronomical instrumentation.

The second theme is the revolution in scientific perspective that spacecraft studies have brought. From the first *Mariner* flyby of Venus in 1962 to *Voyager*'s last encounter with Neptune in 1989, planetary scientists have, time upon time, seen hard-won preconceptions proven embarrassingly simplistic – and sometimes just flat wrong – when spacecraft arrive. In hindsight, the reason for this is obvious: spacecraft observations achieve a resolution and clarity that just isn't approachable from the ground, or even from Earth orbit. The tension between widened scientific perspectives and the history of failed scientific expectations makes one wonder if all of our efforts can get Pluto, the last "astronomer's" planet, right – from afar.

Finally, we wanted to involve the issue of how humans, born and bred on Earth, can come to know exotic and far distant worlds as real places. It is a question of how the mind's eye works. The kings and queens of sixteenth-century Europe once imagined the Spice Islands,

China, and North America. Our generation's Spice Islands are the worlds of the solar system.

If this book succeeds in reaching our goals and in aptly addressing these things, it is in large measure because of the many friends and colleagues who provided advice, critique, practical help, and encouragement along the way. That list includes Fran Bagenal, Kelly Beatty, Rick Binzel, Bonnie Buratti, Steven Brewster, Marc Buie, Steven Dick, Jim Elliot, Robert Farquhar, Don Hassler, Helen Horstman, Garth Hull, Jan Ishee, Hal Levison, Jonathan Lunine, Bob Marcialis, Bill McKinnon, Robert Millis, Tony Reichardt, Steve Shore, Kerry Soileau, John Spencer, Rob Staehle, Carole Stern, Sarah Stern, Mark Sykes, Clyde Tombaugh, Roderick Willstrop and the Library of the Institute of Astronomy at Cambridge (UK), Eliot Young, and Lesley Young. Alan also wishes to thank Larry Trafton and Harlan Smith for interesting him in Pluto in the first place, back in 1979 and 1980.

We are also grateful to Greg Franklin of John Wiley & Sons for his constant encouragement, practical support and patience while we were working on this book, and to Andrew Prince for his dedicated and marvelous editing.

The Pluto–Charon system is a fascinating frontier for astronomy, for astronomers, and for the human spirit. We hope very much that NASA doesn't lose sight of Pluto's scientific value and fascination for the public as it plans its explorations!

Spring 1997   *Alan Stern and Jacqueline Mitton*

Prologue: Encounter!

After traveling over nine years and more than three billion miles to cross the breadth of the solar system, the speeding spacecraft is almost upon its twinned target. The years of anticipation are almost over. There is no time to waste.

If you could only be there; there, alongside New Horizons, *the first robotic explorer to reach the last of the nine planets, you would find yourself in a strange but somehow familiar setting. The vacuum of space on the outer fringes of the planetary system is much as it is at home, between the Earth and Moon: empty, odorless, and crystal clear. So too, the stars still shine, as steadily they do just above the Earth's atmosphere, and in their familiar constellations as well.*

But everything is not the same. The Sun is so distant that it barely shows any disk at all, and so feeble that its light is fainter by 1000 times than at Earth, creating an eternal, almost maddening, dusk – a twilight that cannot warm the heart or body. As a result, even the blackest, most sunlight-absorbing surfaces can barely reach 300 degrees below zero in sunlight.

This wilderness is a place beyond human experience. Fortunately, New Horizons *is right at home here. This is the place for which it was designed; and the day too. Its time to explore new worlds is about to arrive.*

Looking sunward, whole worlds appear as specks, mere flotsam against the emptiness we so simply call "space"; their faint crescents are only dim embers, lost against the Sun's glare.

Looking ahead, however, there is something different. There, against the stars of Sagittarius the Archer is the binary planet Pluto–Charon, moving oh so slowly against the distant stars.

Although Pluto is almost 1600 miles in diameter, the half a million miles that still lie between New Horizons *and this ancient, frozen world reduce it to the size of a pinkish little marble held at arm's length. Pluto's moon Charon, about half the size of Pluto itself, is just two degrees (about the width of your palm) to the right of mother Pluto. In contrast to Pluto's warmer color, Charon presents an almost colorless, ashen reflection of the feeble yellow sunlight that illuminates its surface.*

Pluto and Charon, S. Alan Stern and Jacqueline Mitton
Copyright © 2005 WILEY-VCH Verlag GmbH & Co. KGaA, Weinheim
ISBN: 3-527-40556-9

The Pluto–Charon system is remarkable among the planets. It is the only real example of a double planet system ever found. Pluto itself is the only planet with an atmosphere that forms and decays each orbital cycle. Further, Pluto loses its atmosphere to space at a far faster rate than any other planet, as if it were a comet on a planetary scale. Pluto also is apparently the rockiest planetary body in the outer solar system, yet it contains a substantial supply of the lightest common icy substance found in the deep outer solar system – methane. And amazingly, the snows at the feet of Pluto's mountains, and on the floors of its basins, consist primarily of some of the same stuff you inhale with every breath – nitrogen. Most remarkably though, if the theory of Charon's origin is correct, then the Pluto–Charon binary is also the best analog we have for understanding the formation of the Earth–Moon pair.

Even more dramatically, as it became clear in the 1990s, Pluto and Charon are not odd misfits among the giant planets, but are instead the first and most easily detected examples of a large population of small, ice-dwarf planets that were formed in the ancient outer solar system. As such, Pluto and Charon represent more than just a frontier: they are now understood to be a fundamental link between the classic planets and the myriad small bodies orbiting the Sun beyond Neptune. And among the nine classic planets of our solar system, only Pluto remains unexplored.

Now, after a journey of almost a decade, New Horizons is almost upon its goal, and the anticipation is palpable. The robot ship's human controllers pace, and worry. They try to help their baby along, but they know that the events they monitor back on terra firma took place almost five hours earlier. Were word of a problem to reach the warm Earth, three billion miles sunward, it would be too late to do anything about it. The spacecraft is on its own.

New Horizons was born from the ashes of an earlier generation of planetary exploration, when NASA once sent houseboat-sized probes to the outer planets. The slow and heavy precursors to it often weighed in at many thousands of pounds each. So heavy were the old ships that even the largest launch vehicles could not shove them hard enough to make the crossings to Jupiter and Saturn in less than six or eight years. By contrast, nimble New Horizons flashed past Jupiter's orbit in barely 14 months. Good thing, too: the distance from Earth to Jupiter is barely 10% of the way to Pluto.

Though spare, the 1000-pound probe boasts greater sophistication than its successful but Sumo-like predecessors. Aboard the craft are delicate thrusters for precise pointing and nimble course control, computers possessing state-of-the-art, autonomous, "thinking" software, and a telecommunications system capable of receiving the faint, diluted signals sent up from Earth to direct it. Even more impressively, within its belly, the craft packs imaging cameras, spectrometers, an ultra-stable radio beacon, and other sensors capable of probing Pluto and Charon with ten times the precision and perhaps twice the resolution with which the massive Galileo spacecraft probed Jupiter in the late 1990s. They say amazing things come in small packages. Indeed.

Speed is the key that unlocks the universe. And for New Horizons speed was a gift given to it by its Atlas launch vehicle, which, fully assembled, stood 200 feet tall. New Horizons looked like a tiny hood ornament atop the long orange and white javelin that set it coursing across the great expanse of our solar system.

Then, after the rocket's mad 50-minute push from Earth, there was the journey: over three thousand days through the endless, eternal vacuum that knows no ebb and flow, no seasons, no emotion – only silent running. First, through the asteroidal shards that remain of what might have become the fifth planet, then past Jupiter to gain yet more speed, and then off into the dark abyss where Saturn, Uranus, and Neptune reign.

Speeding, cruising, without sound (or even whisper), the New Horizons spacecraft made its crossing. And as it arced out of the planetary system, the constant tug of the Sun's gravity made the journey to Pluto a long, uphill coast.

In Maryland, Colorado, and Texas, the builders of this craft waited. Years passed. Their children grew tall. And still the builders waited. Time passed. So did presidents, and styles, and hot stocks, and cool trends. Earth raced round the Sun almost ten times, heading toward her five-billionth birthday.

As is the nature of journeys across the great deep of interplanetary space, nothing much happened, even at 35 000 or 40 000 miles per hour. Why? Because the solar system is enormous beyond human comprehension. We can say we understand the scale billions of miles, but does it have meaning?

So, despite the numbing speed, the trip was timeless. Astronaut Alan Bean once remarked that the strangest part of his supersonic, three-day dash across the 240 000-mile gap between the Earth and Moon was "that you never pass anything." Now imagine the six-billion-kilometer crossing to Pluto!

Along the way to Pluto there was the occasional hit by a microscopic dust grain adrift from some comet or Kuiper Belt Object. And there were annual spacecraft checkouts and every-so-often course-correction burns commanded from Earth. But basically nothing happened. The main activity of the journey was simply to take good care of the craft and its cargo of sensors, for show time at Pluto–Charon.

Now, however, the trip is almost done. New Horizons is approaching its target 15 times faster than the fastest supersonic fighter ever flew. From the spaceship, now barely hours from its goal, Pluto appears perhaps half the size of the full Moon. Sharp provinces with subtle shades of yellow and ruddy pink abound on its surface, and a thin haze lies over its rounded limb. The telescopes aboard New Horizons can already see craters, mountains, and long, linear ridges on this outpost of the cryogenic outer limits. Even Charon is beginning to look like a three-dimensional globe; and something is strongly glinting on its morning limb ...

Until a spacecraft actually reaches Pluto, we will never know what that something might be, glinting on Charon's sunlit limb. Indeed, as we write this book, such a scenario remains only a dream, for *New Horizons*, though built, has not yet been launched.

Still, in the pages that follow, we can explore the Pluto–Charon binary to the best extent possible from the vantage point of the early 2000s, and the information gleaned over the 75 years of astronomical research since Pluto's discovery.

To do so, however, we must first travel in a different direction. Rather than upward, toward the worlds beyond the sky, we must first travel backward in time, toward the dawn of the twentieth century. And so we begin in a place that is much farther away than Pluto and far less accessible as well – on a frosty Arizona hilltop, in February of 1930.

1
New Frontier

*"Ever since celestial mechanics in the skillful hands of Leverrier and
Adams led to the world-amazed discovery of Neptune, a belief has
existed begotten of that success that still other planets lay beyond,
only waiting to be found."* – Percival Lowell, 1915

A Planet Hunt

The faint starry image jumped a little – very little, in fact. But the
heart of young Clyde Tombaugh, who saw it twitch late that Tuesday
afternoon in February of 1930, jumped a good bit more. Tombaugh
had been hired by Lowell Observatory, in Flagstaff, Arizona, about a
year before. His job: to take up the reins of long-dead Percival Lowell's quest for that romantic denizen of the deep outer solar system:
Planet X. Now, only months into the search, 24-year-old Tombaugh
had found it.

Who was this cherubic planet finder? He was a Midwestern American, born in Streator, Illinois, in February of 1906, the eldest of the
six children his parents Muron and Adella produced.

Clyde had become interested in astronomy early in his childhood.
This interest was fanned when he was still a boy by the loan of a 3-
inch (7-centimeter) diameter telescope from his uncle Lee. Just after
Clyde reached age 16, in mid-1922, his parents moved the family to a
rented wheat and corn farm in Burdett, Kansas. In high school there,
Clyde played track and field, "dabbled in Latin," and played football
with his friends on weekend afternoons. His passion, however, was
astronomy, and his classmates knew it. In fact, his fellow Burdett
High School seniors of 1925 wrote in their yearbook that Clyde had
his head "in the stars."

Too poor to afford college, he took the job at Lowell Observatory
equipped with little besides a high school education, and a fire inside
himself to become a professional astronomer.

The job at Lowell was young Tombaugh's first experience away
from his family. It was, he told us shortly before his death in 1997,

Pluto and Charon, S. Alan Stern and Jacqueline Mitton
Copyright © 2005 WILEY-VCH Verlag GmbH & Co. KGaA, Weinheim
ISBN: 3-527-40556-9

Fig. 1.1: Young Clyde Tombaugh entering the dome of the 13-inch telescope at the Lowell Observatory in the early 1930s. He is carrying one of the large photographic plate-holders for the telescope. (Lowell Observatory photograph)

a little scary to go so far from home. There, far away from friends and family, Clyde was laboriously to photograph the sky and inspect the images in search of astronomer-aristocrat Percival Lowell's last, some thought chimerical, obsession, this Planet X (Figure 1.1). He did not care much that it was tough, almost thankless work. And never mind that there was no assurance of ultimate success. He just dove right in – night after night, week after week, month after month. But his labor *did* pay off, for on a cold February afternoon

Clyde Tombaugh bagged one of the best known astronomical prizes of the young century: a new planet.

The decades-long search for Planet X had been driven by two motivations – one scientific, one more instinctive. First, there had been the mounting observational evidence that some unseen object was tugging at Uranus and Neptune, causing their courses on the sky to differ from predictions. Second, there was the sheer lure of it – the attraction of finding a new world, of making a mark on the immortal annals of discovery, of tilting at Don Quixote's old windmill.

Fig. 1.2: Percival Lowell (1855–1916) in about 1910, founder of the Lowell Observatory at Flagstaff, Arizona. He was the chief motivator behind the observatory's original search for Planet X. (Lowell Observatory photograph)

The search for the ninth planet was largely initiated by Percival Lowell, a wealthy, scientifically literate aristocrat born just before the US Civil War (Figure 1.2). It is a good thing that the instinctive motivation to make the search was so strong in Lowell, because he did not know that the measurements (made by others, mind you), indicating that the positions of Uranus and Neptune were slightly off their predicted tracks, were wrong. Uranus and Neptune were right on course.

But sometimes, ignorance is bliss. And ignorance of the true situation caused Lowell to use the incorrect "residuals" – that is, the difference between the expected course of Uranus and Neptune and what had been measured – to calculate where in the sky the phantom planet should be.

As early as 1905 Lowell organized the first search for his elusive prey, which he called "Planet X." Lowell was not alone in ninth-planet hunting. Several other astronomers pursued the same goal, including MIT-educated William H. Pickering. Dr Pickering called the phantom "Planet O," and organized several searches for it in the early years of the century.

Fig. 1.3: Vesto Slipher (1875–1969) in 1911. After Lowell died in 1916, Slipher became acting director of the Lowell Observatory. Ten years later he was confirmed as director. In 1929 Slipher hired Clyde Tombaugh as an assistant to help in the search for Lowell's predicted ninth planet, then called "Planet X." (Lowell Observatory photograph)

Altogether, at least eight searches were undertaken by Pickering and Lowell between 1905 and 1920. In these attempts, over 1000 photographic images were made – all to no avail. Those were long years of depressingly ill-rewarded work using what now seem depressingly primitive and inefficient tools and techniques. During those long years of fruitless hunting, four presidents passed the American stage, automobiles exploded into the commonplace, the motion picture industry was born, and "The Great War" came (and thankfully went). As the years dragged by, Percival Lowell grew old. Saddened by the world war, exhausted and discouraged, Lowell suffered a stroke and passed away in November 1916. But his dream of discovering a ninth planet would live on.

Fig. 1.4: The Mausoleum at the Lowell Observatory where the remains of Percival Lowell are interred. (Richard Oliver; all rights reserved)

When Lowell died, his will entrusted the money to ensure the continued survival of his observatory and its staff in Flagstaff. "Percy's" will further stated that the search for Planet X should continue, and so it did. His dedicated assistants, led by Vesto M. Slipher, took up the task (Figure 1.3). And just in case the faint-hearted might someday hence waver in their search, Lowell had himself interred in an ornate blue-and-white marble mausoleum less than 50 meters from the main administration building of his former observatory (Figure 1.4). It was just a little reminder, for the staff ...

Entombed reminder or not, the search for Planet X did temporarily cool in Flagstaff after Lowell's death. However, in California, Pickering and his assistant Milton Humason (a man who was to later become one of the most widely respected mid-century astronomical observers) undertook a new search in 1919. But once again nothing was found.

Meanwhile in Arizona, Lowell's widow sued his estate for a significant part of the million dollars he intended for the observatory. The litigation fees drained the operating funds of the observatory, and the search stalled. Nearly a decade passed.

Fig. 1.5: The 13-inch Lawrence Lowell telescope, with which Tombaugh discovered Pluto, as it was originally in its dome at the Lowell Observatory on Mars Hill, Flagstaff. In 1970 the telescope was moved to the Lowell Observatory's site at Anderson Mesa, 12 miles southeast of Flagstaff, but in 1995 it was restored to working order at its original site on Mars Hill. (Lowell Observatory photograph)

Then, in 1925, Lowell's brother Abbott contributed the funds to commission a new tool for hunting Lowell's planet. That tool was a 13-inch (33-centimeter) diameter refracting telescope and camera that together created an unusually large and useful field of view for planet hunting – 164 square degrees. With its combination of broad search swath and good sensitivity, the new telescope would make it worthwhile to renew the search (Figure 1.5).

By 1928, when the new telescope's arrival was imminent, Vesto Slipher had grown older and not so surprisingly had moved into management. His responsibilities had expanded to include the directorship of the observatory. With all the encumbrances that management made on his time, Slipher decided to hire a new technician to assist in the laborious planet search, and he had just the man in mind.

Slipher offered the job to someone who had been enthusiastically corresponding with him and the other astronomers at Lowell for some time. The young man had written to them time after time to share the careful sketches of Mars and Jupiter he had made using a small telescope that he himself had built for $36. His talent and chutzpah together landed him the job. That young man was Clyde W. Tombaugh.

"Young Man, I Am Afraid You Are Wasting Your Time"

Tombaugh arrived in Flagstaff from his Kansas home by train on the wintry afternoon of January 15, 1929. He was away from home for the first time. His father had told him at the station to "work hard, respect his boss, and beware of loose women."

At the station, a graying Vesto Slipher met Clyde Tombaugh for the first time. Tombaugh carried with him a single, heavy trunk, which was laden more heavily with books than clothes. In the letter that offered Tombaugh a job at Lowell, Slipher had not told Tombaugh that he was being brought to Flagstaff to search for Planet X, merely that he would be involved in "photographic work."

It was only on his arrival in Flagstaff that Tombaugh was informed what his job would center on. However, the new planet-finder telescope still was not quite ready. So Tombaugh, astronomer in waiting, was assigned various tasks, ranging from showing visitors around the observatory to shoveling snow and stoking the main building's furnace. It was less than inspiring work, to be sure. Mercifully, however, by early April the new telescope was finally ready, and Tombaugh could begin the search.

Initially, Slipher tutored Tombaugh in observing techniques. Planet X was expected to be at least 15 times fainter than Neptune. They would have to make long, deep exposures of the sky so that the photographs could soak up enough light to detect the prospective planet. Because planets move perceptibly from night to night but stars do not, the key would be to spot a faint, slowly moving target against the fixed backdrop of a myriad stars and galaxies. To do this, each plot of sky would be re-photographed twice over the course of several nights.

Each photographic plate would have to be exposed for an hour or more to detect the faint pinpoint of light for which they were searching, and Clyde would have to adjust precisely the telescope to keep pace with the turning sky throughout each exposure. To ensure the darkest possible sky as an aid to spotting the faint wanderer, the images were to be made around the time of the new Moon.

According to Tombaugh's own account, after a few nights of lessons, Slipher said, "You're doing all right. You're on your own." And so from then on Tombaugh did the photography. Within a few weeks, Slipher also gave Tombaugh the responsibility for inspecting the photographs for evidence of the hoped-for planet. In effect, the search for Planet X now rested entirely with one young man.

Tombaugh told us in 1996 that few people, even astronomers, "have any concept of the grim (plate-to-plate comparison) job that V. M. Slipher gave to me." We agree. Today no one would stand for such a job – a computer would make the comparisons. But the late 1920s was another time, with a wholly different set of technologies to accomplish things. It was a time when the ocean was crossed by steamship, phone connections were made by human operators moving cables across a board, and most iceboxes actually used ice delivered from distant frozen lakes. So too, the 1920s technology for comparing the images with one another for slowly moving points of light was by today's standards frighteningly primitive. Tombaugh's image comparison technology consisted of a machine called a "blink comparator" that allowed him manually to switch viewing back and forth between any two images (Figure 1.6). Stars, which of course remained motionless, could be distinguished from planets and other moving objects because the moving bodies would seem to jump from frame to frame.

When Tombaugh was not up nights photographing the sky, he was at work during the day, developing the plates[1] and inspecting them

Fig. 1.6: Clyde Tombaugh in 1938 at the Zeiss blink comparator, which he used to examine pairs of photographic plates from the 13-inch telescope during his Pluto search. Following the discovery of Pluto, Tombaugh continued this dedicated but ultimately fruitless work in an attempt to find more distant planets. No discoveries were made. The large glass plates Tombaugh used each measured 14 by 17 inches (36 by 43 centimeters). (Lowell Observatory photograph)

for moving objects. It was not easy. It was not glamorous. It was not even very interesting.

Fortunately, Tombaugh's commitment was monumental – and so was his concentration, which he needed in order to combat the sheer drudgery of methodically inspecting hundreds of photographic images. *Each* contained 50 000 to 900 000 stars. Tombaugh was looking to see if one of the faint points of light might move just the right amount from night to night.

Tombaugh would blink plates slowly, methodically, for hours on end. He set out to be a perfectionist about the task, something that demanded nearly Herculean concentration. He later said that he had to take frequent breaks to clear his mind so he could continue concentrating. Tombaugh knew that the penalty for missing the suspected prey was too great to permit his mind to become dulled by the tedium.

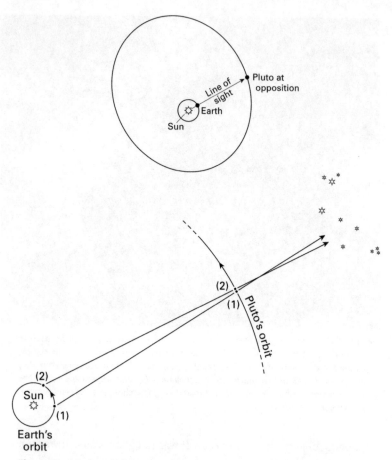

Fig. 1.7: Top: A planet is "at opposition" when it lies in the part of the sky directly away from the Sun as seen from Earth. This occurs when the Sun, Earth, and the planet are in a line in space, as shown here schematically. This is the optimum time for making discovery observations for a number of reasons. One is that the planet reaches its highest point in the sky at midnight, so potential observing time is a maximum. It is also at one of its nearest points of approach to Earth. Another key factor is that a planet's distance can be directly measured at opposition by its rate of travel across the sky. Bottom: Earth travels farther than Pluto does in the time between two observations, (1) and (2). Traveling in its orbit six times faster than Pluto, Earth rapidly overtakes the more distant planet. As a result, Pluto's position in the sky changes relative to the starry background. The change in position over a given period of time is greatest near opposition. The stars themselves do not appear to change their positions between images taken a few days apart because they are more than 10 000 times farther away than Pluto.

After a few months of this work, Tombaugh realized that there is an optimal place on the sky to search for a distant planet. This is around the spot where its motion, as seen from Earth, would be both fastest and most noticeable. At this place on the sky, called the opposition point, a remote planet seems to move westward (i.e., backwards) against the stars (Figure 1.7). Over the course of a year the opposition point[2] sweeps out a great circle on the celestial sphere. This circle is called the ecliptic by astronomers and it passes through the constellations that comprise the zodiac.

A second important advantage of working at the opposition point, Tombaugh realized, was that the amount of motion a faraway planet displays there is inversely proportional to its distance from the Sun. So the farther away an object orbits the Sun, the more slowly it appears to move. Thus, by working at the opposition point Tombaugh could easily distinguish a distant, slowly moving planet from the numerous, nearby, faster moving asteroids that also peppered the images. Tombaugh's realization that the opposition point was the optimal place to focus the search was a crucial advance because it made the detection of truly distant objects so much more straightforward.

Tombaugh presented the opposition-point search plan to Slipher, who approved it. With this new technique in hand, Tombaugh went back to the telescope. Night after night he imaged the sky. He calculated that 30 plates would have to be taken each month to cover the opposition region of the sky as it swept eastward, and he knew that he could only work during those nights each month when the Moon was new or nearly so.

By the time Tombaugh had blinked a month's worth of newly developed plates, the opposition point would have moved on. So he would photograph the next opposition point region when there was no Moon, develop the photographs in the observatory's darkroom, and blink these too. And so forth.

Oftentimes the plates contained microscopic flaws that appeared to create a little jumping point of light when two plates were intercompared, as if to mock him – "Just testing." Almost every plate also contained a handful of those pesky asteroids, which taunted him with false alarms. Fortunately, however, their motion at the opposition point was so large that Tombaugh could tell they were too close to be Planet X itself. After all, he was looking for something that would move just a few millimeters between two plates; the asteroids moved ten times that, or more.

In the course of his searches of the outer solar system, Tombaugh counted over 29 000 galaxies on his plates, plus 1800 variable stars, and discovered two new comets. Months passed, but there was still no sign of Planet X.

By February of 1930, over a year after his arrival in Flagstaff, Tombaugh had worked his way around the sky to the constellation Gemini. He had made a few test plates in this region in early 1929 when the 13-inch telescope was first being set up, but at that time he had not hit upon the idea of using the opposition point to make distant planet detections easier and more systematic. This time as he carefully plowed through Gemini there would be no escape for a stealthy little world. This time Tombaugh ran right into his prey, and caught it.

"That's It!"

It was February 18 when Tombaugh examined his photographs of the star field around Delta Geminorum, the fourth brightest star in the constellation Gemini. These plates had been taken between January 21 and 29. There, a pinprick of light equivalent in brightness to a candle seen from a distance of 300 miles (480 kilometers) jumped ever so slightly from image to image (Figure 1.8).

In the sky there are 15 million stars brighter than the pinprick Tombaugh spied hopping across some of his images in a corner of Gemini, but by blink comparing the position of the anonymous little pinpoint between the various plates, he could see it jumping ever so little!

It did not jump far – only three or four millimeters – but the fact that the jump was so small was the exciting part, for that itself indicated that if the object was real then it surely lay beyond Neptune. "That's it!" he said to himself. But in his logbook, his very own X files, Tombaugh simply wrote, "planet suspect" and the coordinates of the tantalizing speck of light. It was 4 p.m.,[3] and then and there a new planet lost its four-billion-year-old anonymity.

After he finished inspecting the other Gemini plates, good soldier he, Tombaugh set out to confirm whether the faint little nomad was

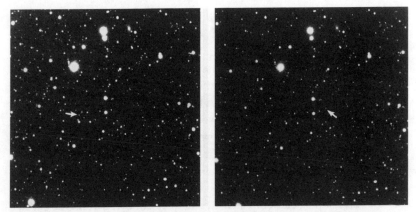

Fig. 1.8: Portions of the original photographic plates on which Tombaugh first detected Pluto by noticing its change in position against the background star pattern. On each plate, the image of Pluto is arrowed. The left plate was taken on January 23, 1930 and the right one six days later, on January 29. (Lowell Observatory photograph)

real. The test he applied was deceptively simple, but stunningly powerful: Tombaugh had taken three other plates of the same region on the same nights, using the smaller, 5-inch telescope that he had bore-sighted onto the 13-inch as a kind of finder scope. Tombaugh's planet-suspect had to appear in the same place on these plates as well if it was real, rather than just an unfortunate artifact of the 13-inch telescope and its camera. And it *was* there.

Bingo! The little wanderer appeared right where it should on all three of these check plates. For nearly three quarters of an hour as he checked and cross-checked the moving pinpoint among the various plates taken, Clyde Tombaugh was the only person in the world to know that Planet X had very likely been found.

Then, sure of himself but trembling with excitement, Tombaugh went to see his boss. Slipher was with Lowell Observatory's assistant director, Carl Lampland. Standing at the door of Slipher's office, Tombaugh knocked and announced directly, "I have found your Planet X."

Tombaugh had never before come to Slipher to make such a statement. He had never set off a false alarm, and both men knew it. On hearing Tombaugh's words, Slipher and Lampland rushed to the blink comparator to check Tombaugh's plates. The two older men confirmed Tombaugh's findings. "Don't tell anyone until we follow it for a few weeks. This could be very hot news," ordered Slipher. Excited, but cautious, Slipher wanted more evidence before going public.

Because three weeks had passed since the Gemini plates had been taken, an even more solid test thus presented itself. Based on the direction and degree of motion from January 21 to 29, a precise prediction of the suspected new position of the X prize could be made. If Tombaugh's target was truly a distant planet, it should show up on new plates about a centimeter from where it had been in late January. This kind of prediction check is a hallmark of careful observational astronomy. It had to be done – and done before any announcement could even be contemplated. It had to be done in part because it was in their own interest, but also, more importantly, because it was in their very bones as astronomers to check, and recheck their discovery so as not to set off a false alarm that could damage reputations.

However, as the sky darkened that evening, clouds covered northern Arizona, preventing planet hunting. The suspense must have been hard, but Tombaugh had no way to settle the matter! He would have to wait until the next night to try again. Frustrated and bored, Tombaugh went into Flagstaff to watch a movie. *The Virginian*, with Gary Cooper, was showing. Word was it was going to be a hit.

Fortunately, the sky was clear on the 19th, and Tombaugh was able to capture another photograph of the little spot of light placed against the pregnant star field. By the next morning, Tombaugh had developed the new plate. Then he, Slipher, and Lampland examined it together. Slipher also brought his brother, Earl, along.

The scene must have been a classic, with four pairs of eyes jockeying for position at the blink comparator. Lo and behold their little IT was there! And IT was exactly where the motion in the earlier images had predicted it would be.

Tombaugh's notebook records the reaction: "Each of the group took a look. There it was, a most unimportant-looking, dim, star-like object which had moved perceptibly … but no disk could be made out."

No disk? Every planet ever seen by human eyes had showed a disk. Like its motion, this was something that distinguished the appearance of a planet from the stars. But maybe this one should not show a disk. It was, after all, very, very faint. Planet X had been predicted to be ten or twenty times fainter than Neptune, but this thing (maybe they should have written it as x, rather than X) was 250 times dimmer than Neptune.

Brother Earl reasoned that it might be small enough, and faint enough, and far enough away just to be a pinpoint, without detectable dimension in their tiny telescopes. To prove his point, Earl built a box with a disk-like hole in one side and a light within, and carried the box a mile from the observatory to convince the group that a faint and far away enough disk would look star-like. As Earl disappeared into the night, ever more distantly, the little disk grew smaller, and smaller, and smaller, until it finally became a pinpoint, just like X! It was a low-tech *tour de force* demonstration that convinced the others: if this new object was sufficiently small and sufficiently far away, it would move but not show a disk.

Now there was beginning to be a feeling that X was really a planet, though perhaps not the large one that Lowell and others had searched for and expected. Ever cautious, Slipher wanted to be even more sure. So, three further weeks of monitoring the motion of the new planet followed, verifying its rate of motion again and again. Amazingly, during all this time, no one breathed a word of it beyond Lowell Observatory.

Then, finally, on the evening of March 12 in Flagstaff,[4] Slipher released the news. After having followed Planet X for seven full weeks, Slipher officially announced the discovery on behalf of Lowell Observatory.

Slipher had selected the official date of March 13 to make the announcement because it held special significance: it was both the 149th anniversary of the date on which Uranus, the first planet[5] to have been discovered telescopically, was initially spotted in 1781, and it would have been Percival Lowell's 75th birthday.

Slipher's first announcement was in the form of a telegram sent via the observatory's trustee (Lowell's grandson, R. L. Putnam) to the Harvard College Observatory, which was the official clearinghouse for discoveries of new asteroids, comets, and novae. Telegrams were typically terse in those days and relatively expensive – one was charged for every word. Slipher's short missive read:

> "Systematic search begun years ago supplementing Low-
> ell's investigations for Trans-Neptunian planet has re-
> vealed object which since seven weeks had in rate of mo-

tion and path consistently conformed to Trans-Neptunian body at approximate distance he assigned."

The honor of a second announcement was given to Lowell's widow, Constance. With Mrs Lowell's announcement the circle was complete.

It was done. The world now knew X existed. And the world responded. By 1930, the global village was already a real (if comparatively quaint) electronic construction. Headlines echoed the news: "Ninth planet Found at Edge of Solar System." "9th Planet, X, Is Found by US Scientists." Tombaugh's parents found out from a reporter! Muron's boy had become the hero of the story and a celebrity.

Within days, Tombaugh and company at Lowell were famous. Today the discovery would probably net the standard Warhol allotment of 15 minutes of fame. But in 1930 fame lasted a little longer. The media rush following the discovery went on for months.

But What to Call It?

Something deep-seated in human nature calls on us to name things. It is almost as if a thing is not real, or whole, until we name it. And so of course X *had* to have a name.

Suggestions flooded in: "Zeus," "Cronos," "Minerva." Widow Lowell herself first liked "Zeus," but then suggested "Percival," then "Lowell," and even "Constance," her own name, as candidates.

Soon dozens of other well-meaning suggestions poured in as well. There were hundreds, then thousands. But when all was said and done, the name that the Lowell staff preferred was the one suggested by 11-year-old Venetia Burney[6] of Oxford, England (Figure 1.9) – "Pluto," the Greek god of the Underworld, the brother of Jupiter, Neptune, and Juno, and third son of Saturn, who was able when he wished to render himself invisible.[7]

On May 1, 1930 Lowell Observatory's director, Vesto Slipher, lent his support to naming the new world Pluto; that clinched it. Both the American Astronomical Society and the UK's Royal Astronomical Society later that month adopted Pluto as the official name and ♇ as the official symbol for the new world. ♇, we note, was Percival

Fig. 1.9: Venetia Burney, who as an 11-year-old English schoolgirl suggested the name "Pluto" for the planet newly discovered by Clyde Tombaugh. (Courtesy Mrs Venetia Phair, née Burney)

Lowell's monogram, something that was not lost on those making the naming decision.

Pathways of the Gods

While the international press heralded the discovery of Pluto, astronomers around the world rushed to work out the orbit of the new planet. This was the natural first order of business. Without an orbit the ninth planet's ever-changing position could not be predicted, and the new world could become lost. Worse, its context among its eight more familiar sisters could not be even crudely judged until its orbit was known.

What exactly is a planetary orbit? The orbits of the planets are, simply put, their tracks around the Sun. Each orbit a planet makes is a lap on this repetitive course, and takes a period of time called the planet's "year."

The basic framework for understanding orbits was laid down in the sixteenth century by the Renaissance astronomer Johannes Kepler. Kepler's discoveries resulted from the mental sweat equity he invested to distil patterns from the motion of the six planets originally known to the ancients (Mercury, Venus, Earth, Mars, Jupiter, and Saturn). His three discoveries are so fundamental that they are remembered today, in his honor, as Kepler's laws. They are so powerful, and so utilitarian, that barring some disaster that sets humankind back to a medieval world, Kepler's laws will continue to be used by astronomers, navigators, and space pilots as long as our species endures.

Kepler's first law states that the orbit of a planet about the Sun is an ellipse (Figure 1.10). Kepler's second law states that the line joining a planet to the Sun sweeps out a constant area per unit of time. Sounds a little thick, perhaps, but from a pragmatic standpoint, rule number two means that planets move around the Sun more slowly when they are far from the Sun, and more rapidly when they are closer to it. Kepler's third law describes a simple equation, a numerical recipe, for calculating the length of a planet's "year" almost exactly, using just two observable quantities: the mass of the Sun and the planet's average distance from it.

In the three and a half centuries between the discovery of Kepler's laws and their application to Pluto, a considerable body of theory and observation had accumulated concerning the inner workings of orbits and the architecture of the solar system. By the late eighteenth century, for example, it had become possible to calculate the orbit of a planet from a few well-separated measurements of its position against the stars, and to then use the orbit solution so obtained to predict with high accuracy where the planet would be on any future date. This is not hard for silicon circuits, but it is not a trivial affair for computers made of flesh and blood. Why? Because in addition to the force exerted by the Sun's gravity, each planet's gravity also slightly affects the motion of the others. These other planetary tugs explain why Kepler's calculations were almost – but not quite – exact.

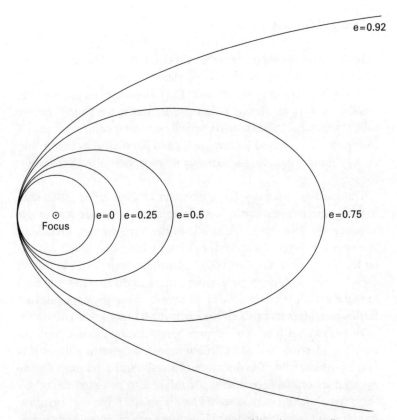

Fig. 1.10: As Johannes Kepler discovered, the orbits of the planets are ellipses, not perfect circles. This diagram shows a circle (zero eccentricity) and four ellipses of different eccentricity. The more elongated an ellipse, the greater its eccentricity, e. The point labeled "Focus" is the center of the circle and also one of the two foci of each ellipse. When a planet travels around the Sun in an elliptical orbit, the Sun is at one focus of the ellipse, which is not the center point (unless the orbit is perfectly circular). The ellipse with eccentricity 0.25 is about the same shape as Pluto's orbit. The ellipse with eccentricity 0.92 is so stretched out that only part of it can be shown. It is similar in size and shape to the very elongated orbits followed by comets such Halley, with periods of many decades or longer.

Against this background, many of the world's astronomers set out in 1930 to learn Pluto's orbit and compute its future path. Before revealing the surprises that they found, however, it is worthwhile to take a little detour to see the forest of the solar system as a whole, before we focus on the solitary sapling called Pluto.

Empty of Empties

In overview, the solar system is laid out with the Sun at its center, and the orbits of the widely spaced planets concentrically arranged outward, from Mercury to Neptune. The four innermost planets – the smaller, rocky ones – move in tight orbits close to the Sun. These so-called "terrestrial" planets share broadly similar qualities of size and rockiness. Apart from Mercury, each has a substantial but razor thin atmosphere surrounding it. Mercury's atmosphere is but a gossamer corona.

The smallest, Mercury, has a diameter of 3050 miles (4880 kilometers); the largest, Earth, has a diameter of 7970 miles (12 753 kilometers). The closer a planet is to the Sun, the more rapidly it completes its orbit. Mercury, for example, lying only about 40% as far from the Sun as the Earth does, completes each circuit in just 88 days.[8] Venus, which is 70% as far out as Earth requires 225 days to make each lap, and the Earth, of course, takes 365 days. Moving farther out, Mars takes about 1.9 Earth years to complete each orbit.

Venus and Earth orbit on ellipses barely distinguishable from circles, but Mercury and Mars follow more exaggerated ellipses that range farther afield. The deviation of an elliptical orbit from a circle is called its orbital eccentricity. (It might also be called the orbit's "eggyness," but this would sound less scientific.) The eccentricity of Earth's orbit is a little under 2%. Venus's orbital eccentricity is less than 1%, but Mars's is near 10% and Mercury's is almost 20%.

Beyond Mars, the final outpost of the terrestrial planets, lies the rocky asteroid belt, and then the vast and unabashedly bizarre wilderness inhabited by the four giant, outer planets: Jupiter, Saturn, Uranus, and Neptune. These Goliaths are enormous, cold worlds with poisonous atmospheres a thousand times deeper than Earth's thin but refreshing blue and breathable shell. Jupiter, the largest of the giants, could fit some 1400 Earths within itself! Even Neptune, the smallest of these giants, could swallow over 60 Earths (Figure 1.11).

The four giant planets travel in orbits that are all circular to within 5%. Their orbital periods (i.e., their own years) range from 12 Earth years for Jupiter to 167 years for Neptune. Jupiter and Saturn are bright enough to see easily by eye, and were known to the ancients as special "stars" that moved across the sky, like Mercury, Venus, and Mars. By contrast, Uranus and Neptune are much fainter, and were

Fig. 1.11: The relative sizes of the nine major planets and the Sun are shown here to scale.

discovered only after the invention of the telescope and its application to astronomy.

The orbits of all eight planets from Mercury to Neptune lie close to a mathematical plane, an imaginary flat sheet extending out into space near the Sun's equator, which astronomers call the "invariable plane." The four inner planets all orbit within 6 degrees of the invariable plane. The four giant planets orbit even closer to this plane – none strays from it by more than 2 degrees.

Now consider the accompaniments of the planets: their rings and moons. Here, there is another major difference between the inner and the outer planets. The inner planets are not encircled with rings, and have *in toto* just one large satellite (our Luna) and two tiny moons hardly larger than a respectable county (Mars's Phobos and Deimos). In contrast, each of the giant planets displays a system of rings, and each possesses dozens of moons. Among the more than 100 moons discovered around the giant planets to date, five are larger than our own moon, and three are larger than Mercury itself. Most, however, are tiny, like Phobos and Deimos.

When we step back from these details, these trees that form our solar system's forest, something even more fundamental emerges. The central, overarching aspect of the planetary system that we all so commonly casually overlook is that it is almost completely empty.

The planets themselves are but specks, spaced across millions and even billions of miles of nothingness. If the Earth were reduced in size to a simple, blue basketball, the Sun itself thus illustrated would be just 100 feet (30 meters) in diameter. In this scale model, the Sun would lie almost 2.5 miles (4 kilometers) away from basketball Earth, with nothing but the little basketball Venus and baseball Mercury occupying the space between the two. In this miniature solar system, Jupiter, about 12 feet (3.7 meters) in diameter, would circle the Sun 10 miles (16 kilometers) away, followed by Saturn 20 miles (30 kilometers) out, Uranus some 40 miles (64 kilometers) out, and lonely Neptune 60 miles (97 kilometers) distant from the tiny yellow hearth, glowing down below. Thus, within the 80 miles (129 kilometers) corralled inside Neptune's orbit in our Tinker-toy™ model, there is nothing but the eight planets known prior to Pluto, their tinier-still retinue of satellites, together with a few thousand scale model asteroids (most no larger than sand grains), and perhaps a few hundred comets, each, like the asteroids, barely a grit of sand.

Empty of Empties! The magnitude of the solar system's emptiness is the essence of the place. Indeed it is almost but not quite, nothing but emptiness. This, of course, is why we call it "space."

It's a Rogue!

Engulfed within the wild, black emptiness, yet another billion miles in the distance beyond Neptune (20 miles (32 kilometers) in our scale model), Clyde Tombaugh had discovered Lowell's planet; our little *dyevoshka*, our Pluto.

Once Pluto's discovery and position became public knowledge, astronomers across the Americas, and Europe, and Asia, and Africa began to collect frequent observations of its position, so that an accurate orbit could be determined.

Two types of positional data came in. Most came in the form of measurements from new photographs, but some also came from the astronomical archives of the world, as various groups recognized that they had unwittingly photographed patches of sky containing Pluto over the 15 or so years prior to 1930. This kind of research is now common in astronomy, thanks to computer databases, but in the 1930s it was relatively novel since computers did not exist.

The pre-discovery observations of Pluto found in astronomical libraries were particularly useful, because they greatly extended the timebase over which accurate positions were known. Once located, they rapidly gave rise to an accurate orbit.

By the early summer of 1930, some 136 archival images of Pluto had been located on old plates taken before Tombaugh's discovery. Ironically, Seth Nicholson found that Milton Humason's old 1919 search for Pickering's Planet O had in fact netted four images in which Pluto faintly appeared! But Humason had not recognized the planet, in part because he had imaged it away from the opposition point, where planet detection is easiest and least ambiguous.[9]

Armed with accurate positions stretching back as far as 1914, several astronomers calculated and published similar solutions for Pluto's orbit in 1930 and 1931 (Figure 1.12). As a result, within about a year after Pluto had been found, its orbit had been reliably established. This is what was found:

- Pluto takes 248 years to complete an orbit – a record among the known planets. When Pluto was discovered it was heading toward its closest approach to the Sun, called perihelion, which it later reached in 1989. Its previous two perihelia occurred in 1741, when George Washington was only a boy, and in 1493,

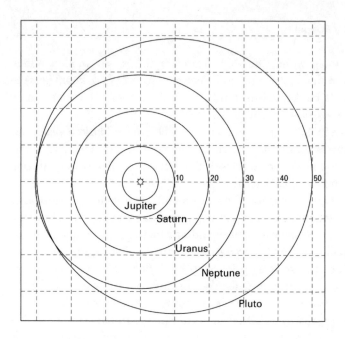

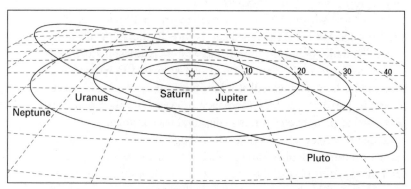

Fig. 1.12: The layout of planetary orbits in the outer solar system. The grids are labeled in astronomical units. Looking down on the plane of the solar system (top) shows how Pluto's strongly elliptical orbit crosses inside Neptune's. The foreshortened perspective view (bottom) shows the inclination of Pluto's orbit relative to the orbits of the other outer planets. The orbits of the inner planets, including Earth, are not shown here.

just months after Columbus had reported his explorations to the Reyes Catolicos. Pluto follows a long orbit indeed!

- While the other planets of the outer solar system follow ellipses that are typically eccentric by only 1% or 2%, Pluto's orbit stretches ellipticity to a new record among all the known planets – 25%! Each orbit Pluto wanders over a range of heliocentric (i.e., solar-centered) distances – from 3 to 5 billion miles (5 to 8 billion kilometers).
- Whereas all the other planets orbit within a few degrees of the invariable plane, Pluto's orbit is tipped away 16 degrees, far askew from the others.

Together, these findings painted a picture of a nearly fantastical orbit, unlike that of any other planet. Pluto's path is so strange, in fact, that near perihelion it actually crosses the orbit of Neptune, making it temporarily planet eight (not nine!) for a few decades every two centuries.

No other planet crosses inside another's orbit. How dare planet Pluto do that! Why? What kind of thing was this rogue roaming the frigid outback of the planetary system? It would take the remainder of the twentieth century to find out.

2
First Facts

"Every book of descriptive astronomy is to-day out of date."
– A. C. D. Crommelin, President of the Royal Astronomical Society,
on the announcement of the discovery of Pluto, March 13, 1930

For almost 50 years after its discovery, across the great middle of the twentieth century, even the most basic outlines of Pluto's true nature remained the secret they had always been. The technology available to astronomy was simply too primitive to penetrate the distance between us and that dim, faintly reflecting world. Through the 1930s to the 1960s little was revealed. By the 1970s, however, the availability of more modern instrumentation had begun to reveal the face and the persona of the solar system's ninth planet.

Most of the story told in this book concerns the rapid unwrapping of the scientific gift that Pluto has revealed itself to be. But before we tell the Tofflerian tale of the accelerated, almost madcap pace of Pluto discoveries that rolled across the research landscape of planetary astronomy in the 1980s and 1990s, it is worth returning to the yesteryears of the mid-twentieth century, when the hard-won, first clues to Pluto's nature were revealed.

The Dark Ages

Despite the fact that the outer planets are tens of thousands of times nearer to us than the closest stars, it is often harder to make progress in studying such nearby worlds than it is in stellar and galactic astronomy.

Why is that? For one thing, there are fewer bodies to study, and each planet is distinctly unique. As a result, it is harder to find trends among the observations and come to general conclusions. For another thing, the number of planetary researchers and the resources available to them are smaller. But most importantly, planets do not cooperate to the same degree that stars do. Stars shine, by virtue of their

Pluto and Charon, S. Alan Stern and Jacqueline Mitton
Copyright © 2005 WILEY-VCH Verlag GmbH & Co. KGaA, Weinheim
ISBN: 3-527-40556-9

internal energy sources, as do galaxies and glowing nebulae. And the light that stars and galaxies emit reveals intimate details about their nature and inner workings. By contrast, planets reflect light from the Sun, but for practical purposes they emit no light of their own. Thus, they are more impervious to the tools of observational astronomy.

Pluto, being the farthest and the faintest planet, is *the* challenge of the planetary system. Why? For one thing, Pluto's maximum apparent diameter as seen from Earth is hundreds of times smaller than that of either Mars or Jupiter. To image Pluto's surface requires resolving something as tiny as a walnut 30 miles (50 kilometers) away!

As difficult as that may sound, the job of studying Pluto is harder than just cracking the nut that distance imposes. Pluto at its brightest is also 14 *million* times dimmer than Venus is at its brightest, and a million times dimmer than Jupiter's best; indeed, Pluto is even 1500 times fainter than Jupiter's major satellites, Io, Europa, Ganymede, and Callisto, which were not well understood until the Voyager spacecraft reached them in 1979.

So what was the point of struggling to learn about Pluto through the long years from the 1930s to the late 1970s? It was the challenge and the attraction of exploring a truly frontier world.

In those five decades of the middle of the century, when many of the leaders of astronomy today were either still not born or in the awkward stages of early life, it was just barely feasible to learn how long Pluto's day lasted, and the color of its surface. Later, in the 1970s, it was possible to find out that it lies tipped on its side, as Uranus does, and to discover something important about Pluto's surface composition. It was even possible to discover that Pluto has a companion, in the form of its large moon Charon. These meager gains revealed the first sketchy details of the Pluto we know today. So too, they illustrate the budding of planetary astronomy that occurred in the middle of the last century. It is a story worth exploring in some detail ...

Shimmer, Little Planet

Pluto's small size, great distance, and faintness together conspired mightily against all efforts to obtain even a crude characterization of it for 60 years. One reason for this is that all our telescopes suffered a common, crippling handicap: they were at the bottom of Earth's turbulent and shimmering ocean of air.

The British physicist Lord Rayleigh discovered in the early nineteenth century that the physics of optical instruments works to ensure that, all else being equal, proportionately larger instruments give proportionately better resolution of fine details. As such, for example, the 100-inch reflecting telescope at Mount Wilson in southern California should provide about seven times the resolution of Slipher and Tombaugh's 13-inch Pluto finder. And in fact it does. The only problem is that the light from Pluto, having traveled 3 billion miles (5 billion kilometers) or more to reach the top of the Earth's atmosphere, becomes defocused in that last, nasty 50 miles (80 kilometers) or so it must traverse across the thin arc of Earth's atmosphere on its way to the telescope. Astronomers often refer to this defocusing effect as "seeing." Seeing can vary from moment to moment, from place to place, and even along different lines of sight through the sky, depending on the amount of turbulence in the overlying atmosphere.

You might think that atmospheric seeing is not a large effect. After all, it only limits the resolution of a telescope to between 1/7000 and 1/1000 of a degree, or, as astronomers say, between 0.5 and 3 arcseconds (an arcsecond is 1/3600 of a degree.) It does not sound like much, and your eye cannot usually detect it, but if you turn a telescope toward the surface of the Moon, for example, the shimmering becomes immediately apparent.

For Pluto, whose tiny size and great distance conspire to give it an apparent diameter of only about one-tenth of an arcsecond as seen from Earth (1/36 000 of a degree, ten times smaller than the seeing blur due to the atmosphere!), the result is disastrous. Because you had to make time exposures during which the seeing would completely blur Pluto's image, it did not matter whether you used the 13-inch Pluto finder at Lowell or the 100-inch at Mount Wilson. Even using a telescope as large as the present-day 10-meter (400-inch) Keck telescopes atop Mauna Kea in Hawaii would have made no difference: the exposures would still have had to last many seconds using photographic technology, the atmosphere would have prevented Pluto from being resolved, and that means you would not have seen any details on its surface.

Today we can beat this problem by using a combination of fast, digital cameras, sophisticated computers, lasers, and optical correctors to measure and then eliminate most of the atmospheric twinkling. We can also beat the seeing problem now by putting our telescopes into orbit, above the atmosphere, as in the case of the Hubble Space Telescope. But those options were not practical until 60 years after Pluto's discovery. As a result, it was not possible for any telescope, regardless of the quality of its optics or its exquisite guiding system, to reveal a picture of Pluto like the ones of the closer, brighter, and larger disks of other planets, such as Mars, Venus, and Jupiter.

That limitation greatly delayed studies of the ninth planet: no picture no story – well, almost. How, for example, without resolving the disk of a planet, can one map its surface or tell if clouds might be in its atmosphere? Without features to track, how would one even determine the length of the planet's day? Fortunately, astronomers developed a kind of poor-man's substitute for direct imaging that could shed light on some of these very basic questions. The technique is called lightcurve analysis.

The Pulse of a Planet's Brightness

A lightcurve is a graph of how the brightness of a star or planet varies with time. One makes a lightcurve by repeatedly measuring the brightness of an object, say every few minutes or hours, and then plotting up the readings. If the resulting graph is analyzed properly, it is possible to determine a number of useful things, like the length of a planet's rotation period (i.e., its day), the tilt of its rotation axis, whether the planet has complex weather, and just how contrasty (i.e., how varied) its surface might be.

However, this is easier said than done because, although some stars vary by factors of 10, or even 100, on regular intervals, most planets vary in brightness by only a few percent as they rotate. This means that the brightness measurements have to be highly precise. Additionally, they have to be either made under the same circumstances of weather and atmospheric turbulence, or carefully calibrated in comparison to an unchanging star. Such a comparison works because the atmosphere affects both astronomical objects in the same way. Thus, if one chooses a star that is known to have constant brightness, it can be used to compensate for the vagaries introduced by the Earth's atmosphere, which astronomers must tolerate whenever they observe. More subtle complications also arise but the general idea is as we just described it.

How exactly is the brightness of the planet and the comparison star to be measured? One well-established technique for obtaining lightcurves developed around photography, which became a standard data-recording medium for astronomy as early as the 1850s and 1860s. One key advantage of photography over visual records is that it is less subjective. A second important advantage is that a photograph can be analyzed by different astronomers to cross-check the results. Most importantly, however, a photograph can measure the brightness of fainter objects than the eye can see. How? By leaving the shutter open for a period of time and exposing the image "deeply."

Without any doubt, photography brought about a scientific revolution in astronomical research in the century between 1850 and 1950. It allowed the science of astronomical observing to move from sketches and drawings, made by humans translating what they saw at the back end of a telescope, to a reliable, repeatable, easily reproducible product that could be *quantitatively* analyzed to reveal faint subtleties no eyeballed sketch could show. Among its many successes, photography made possible the mapping of faint nebulae, the recording of spectra for chemical analysis, and the discovery of a myriad asteroids, and of course, the discovery of distant, slowly moving Pluto.

By the mid-twentieth century, there were many specialized photographic emulsions for astronomical use, but all astronomical photography suffered a common weakness, called "nonlinearity." What a great word, "nonlinearity." Toss it out at a party sometime just to see who leaves and who stays in the conversation ...

What astronomers mean by this piece of jargon is actually straightforward: if two objects x times different in brightness are imaged onto the same photograph, it is unlikely that their resulting images will be precisely x times different in brightness. They may be close, but they will not be *exactly* proportional, or linear. As a result, accurate, quantitative comparisons become difficult.

Nonlinearity also rears its head another way, when the same object is exposed at different times on different emulsions, or even with the same kind of emulsion but with different exposure times or different camera temperatures. Nonlinearity is a plague that infects lightcurve (and other) quantitative studies that use photography, because the uncertainty in the measure of a planet's brightness on a photograph can be as large or larger than the lightcurve of the planet itself. These and other difficulties of working with photographs prevented anyone from obtaining a useful lightcurve of Pluto in the 1930s and 1940s.

Although photography revolutionized astronomical capabilities a century ago, very few astronomers today ever actually deal with photographs because photography has been almost entirely replaced by a far more accurate technique using compact solid-state digital cameras, which are described in Chapter 3.

Unfortunately, when Pluto was discovered in 1930 the modern-day solid-state devices we take for granted today did not exist. However, by 1950, the electronics revolution was beginning, and an early device called a photomultiplier tube (or PMT) was becoming widely available. Because of their advantages over photography, PMTs were popular for making accurate light level measurements, which astronomers call "photometry."

PMTs are not as sophisticated or useful as cameras: They are just simple light meters that do not reveal an image of their subject – only a measure of its total brightness. But the good news about PMTs is that they produce far more linear results than photographic emulsions. An x times brighter object gives an x times brighter signal. What an improvement! (What a relief!)

Furthermore, because they amplify the signal they detect many times, PMTs can be much more sensitive than photographs. The bad news is that PMTs are large and bulky, and only record the total light falling on them. So much so, that it is usually the case that just one PMT fits at the place where the camera goes at the back end of the telescope. As a result, instead of getting a two-dimensional image with thousands of resolution cells that a photograph gives, a PMT gives you just one pixel – one picture element. There is no image, just a single measure of the brightness of the thing you're looking at. Point it at Jupiter and you get the total brightness of Jupiter. Point it at the Moon and you get the brightness of the little plot of land on the Moon the telescope sees, but not an image of that spot, just its total brightness (Figure 2.1).

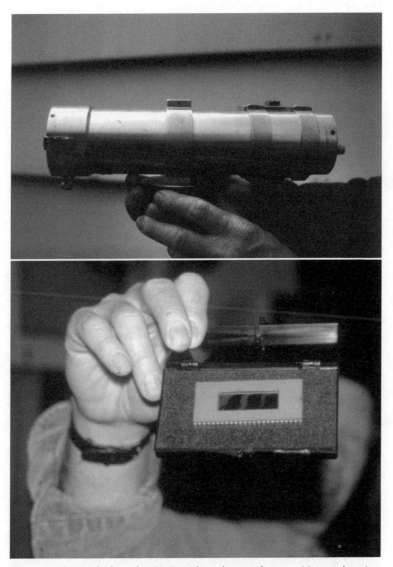

Fig. 2.1: A photomultiplier tube (PMT) such as the one shown at (a), seen here in its metal housing, is capable of recording just one item (or pixel) of data – the total intensity of all the light falling on it. The main advantage of a PMT over the analysis of the brightness of an object on a photograph is the PMT's greater accuracy under most circumstances. By comparison, the smaller charge-coupled device (CCD) (b) records the intensity of light on each of the many thousands or even millions of individual tiny pixels that make it up. In this way, a CCD can produce a complete two-dimensional image, similar to a photograph, but it records the variations in the intensity of light across the image much more accurately than a photograph does.

Now, back to our lightcurve story. All we need to form a lightcurve is an *accurate* record of how the brightness of the whole of Pluto changes over time, and that is what a PMT can provide.

So when PMTs became available to astronomy in the early 1950s, Bob Hardie of Vanderbilt University began a project to measure Pluto's lightcurve. If he succeeded, he knew he would obtain the first physical information on Pluto itself since its discovery, for up to that time nothing was known about the planet other than its orbit and its approximate color.

Two early attempts on Hardie's part in 1953 resulted in good evidence that Pluto varies, but that was all. Nonetheless this was an encouraging sign, since many planets, including Mercury, Venus, Uranus, and Neptune, do not show strong variations in brightness – just completely flat lightcurves.

In 1954 Hardie was joined by colleague Merle Walker and together they set out to improve on Hardie's earlier results. They first observed Pluto together early in 1954 and then, in the spring of 1955, the pair secured observing time on the largest telescope at Lowell, a 42-inch reflector. As they collected their data, Walker and Hardie reduced it to a common standard using two stars as their reference posts. Within a few weeks it became clear that Pluto was much more variable than most planets: indeed, its brightness fluctuated by more than 10%. Even more surprisingly, Pluto's brightness fluctuations precisely repeated themselves like clockwork, with a period of 6.39 days, plus or minus a mere 4 minutes. A decade later Bob Hardie wrote a popular account of these findings in *Sky & Telescope*:

> "During each rotation, the planet's light increases for about four days, then drops to a minimum again in about two. If Pluto is assumed to have a more or less round disk, evenly illuminated out to its edge, the lightcurve asymmetry can be explained only by an extremely unlikely pattern of dark and light surface markings."

The information Hardie and Walker obtained, though meager, was revealing indeed. Clearly, Pluto's variability was remarkably large, larger in fact than for any of the other planets except Mars. As the *Sky & Telescope* article revealed, even Bob Hardie found it hard to believe how stark the surface markings implied by the lightcurve must be.

Pluto's metronomic repetition also meant that, almost without doubt, the surface of Pluto was being seen through little if any atmosphere. Certainly, if there were an atmosphere, then it did not include lots of highly variable clouds, for if it did, the lightcurve would have displayed an ever-changing set of bumps and wiggles on it. Further, the shape of Pluto's lightcurve was studded with little hills and dales that hinted at a potentially complex set of surface markings just waiting to be revealed, if only Pluto could be imaged. Finally, the lightcurve's long period, 6 days and 9.4 hours, was itself remarkable: no planet beyond the orbit of the Earth except Pluto takes more than 25 hours to rotate. Indeed, the gas giants that dominate the outer solar system each spin round with whirlwind days ranging from 10 to 20 hours. Why, astronomers wondered, did Pluto rotate so slowly?

These facts carried the impact of an artillery shell landing among the planetary scientists of the day. Here, on the ragged outer edge of the planetary system was a world on a bizarrely different orbit than any other, rotating at glacial pace, and possessing either little atmosphere at all or, at least, a nearly cloud-free one (else the lightcurve would vary due to cloud patterns). These are not the attributes of the other outer planets, which are large and rapidly rotating, with deep, cloudy atmospheres. Pluto was already revealing itself to be very, very different from any of the planets lying closer to the Sun.

A Cockeyed World

The discovery of Pluto's strong lightcurve was an unexpected and fascinating find. It spurred forward efforts to apply newer and more difficult types of measurements to faint Pluto, and it encouraged researchers to improve the lightcurve data in an effort to yield further results.

New lightcurves were obtained in 1964 by Hardie himself, and then in 1972 by Leif Andersson and John Fix. From these it was discovered that Pluto's lightcurve was changing in two ways. First, the average brightness was gradually falling, which indicated that Pluto was dimming (even though it was moving closer to the Sun). Second, the amplitude of Pluto's lightcurve – that is, how much it varies from its peak to its minimum – was increasing with time. The decrease in Pluto's overall reflectivity was a small effect, but the

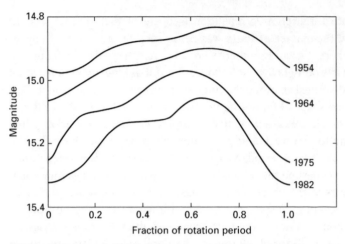

Fig. 2.2: The change in Pluto's lightcurve over almost 30 years, from 1954 to 1982. During this time, Pluto became dimmer and its variation in brightness over its 6.4-day rotation cycle became more extreme. (Brightness is shown here on the magnitude scale, in which larger numbers mean fainter.) To ensure a fair comparison, these four lightcurves have been adjusted to correspond to Pluto being illuminated by the Sun in exactly the same way and at the same distance from the Sun. This adjustment proves that the observed change over time is the consequence of a real change in the brightness of the visible surface of Pluto. (Adapted from S. A. Stern et al. 1988, *Icarus* 75)

increase in lightcurve amplitude was not: it doubled between 1954 and 1972 (Figure 2.2).

Bob Hardie conjectured that Pluto's surface might be changing its reflectivity due to the movement of surface frosts as the planet approached perihelion. After all, Pluto had moved almost a billion miles sunward since its discovery. But was not this exactly the opposite effect that common sense might suggest? Should not Pluto grow brighter as it moved closer to the Sun? It was puzzling indeed.

Andersson and Fix proposed an interpretation that explained both the change in average surface brightness and the increase in lightcurve amplitude. Because the lightcurve seemed to maintain its gross shape even as the difference between the minimum and maximum brightness increased, they reasoned that the tilt of Pluto's axis was probably large. If that were in fact so, then it would cause different bands of latitude to be visible at different times as Pluto moved around the Sun and was seen from different perspectives. The tilt of a planet's spin axis relative to its orbital plane is called its obliquity.

Earth's tilt, or obliquity, is 23.5 degrees. It is just this tilt that generates our seasons. Among the planets from Mercury to Neptune, obliquities range from Mercury's nearly 0 degrees, to Uranus's 98 degrees. In fact, except for Uranus, which is literally lying on its side, all of the planets rotate with a tilt of 30 degrees or less. Like Uranus's, Pluto's obliquity seemed to be very large.

Andersson and Fix constructed a computer model to see what values of the obliquity could explain the trend of varying lightcurve shape. They discovered that Pluto's obliquity had to be large – at least 50 degrees – and perhaps quite a bit higher still. That is quite some tilt!

Further, they concluded that the angle from which Earth was viewing Pluto, called its aspect angle, had moved equator-ward from 1954 to 1972. The decreasing brightness of the planet over this 18-year time span indicated that Pluto's polar regions are brighter than its equatorial regions.

Did this mean that there are polar caps on Pluto? Perhaps – but the case was pretty meager, and the evidence very indirect. It was also possible that Pluto's appearance had changed for other reasons, such as some kind of actual change in the patterns of markings on the surface. To determine if there were in fact polar caps would require more and better evidence. After all, at the time it was not even clear whether there were ices *anywhere* on the surface. To search for ices it was necessary to explore Pluto's surface composition, which calls for a technique called "spectroscopy."

Reading the Rainbow

Thanks to color vision, all of us who can see and are not color-blind carry through life a crude but effective spectroscopic capability within each eye. Our eyes recognize that light comes in a range of wavelengths, which it displays in a delightful panoply of reds, yellows, blues, and all the combinations in between. Color perception, like our other senses, continuously conveys important information to us, and, at the same time, adds a delightful richness and esthetic quality to our lives. Given the experience of color vision so deeply rooted within us, it is not surprising that astronomers invented a way to measure precisely the colors of the distant stars and planets as a means to understand them.

The basic technique invented to measure colors is easy: simply make two images of the object of your interest using filters that pass different colors – for example, red and blue. If the object is brighter in the red image, then we say its color is red, since it must be emitting or reflecting more red light than blue light. Similarly, if the bluer filter gives the brighter image, then we call the object blue. If the two filters yield equally bright images, however, then the object does not have much intrinsic color, and astronomers call it gray.

The idea is very straightforward. In practice, however, color measurements have some subtle idiosyncrasies that must be carefully taken into account. For example, different color filters transmit light with different efficiency. So too, the detector used to record the light usually has better sensitivity at one end of the spectrum (either red or blue) than the other, depending on its type. If, for example, the detector is most sensitive to blue light, then it might register a larger signal through a blue filter than a red filter, even though the object's true blue brightness is weaker than the its red brightness. These and other factors make the game a bit complicated, but then part of the fun in this is knowing the tricks of the trade and executing them flawlessly. For example, there are tables of numbers prepared for each filter to correct for their individual efficiencies; it is also possible to calibrate detectors, or even the whole observing system, by looking at a test pattern or distant object with well-established, "reference" colors.

In the very month that Pluto was discovered back in 1930, Carl Lampland compared the ratio of Pluto's brightness through yellow and blue filters to the ratio he measured using the same filters for Neptune. This type of comparative or "relative," measurement is self-calibrating, in that all of the hard parts of the calibration turn out to be the same for Pluto and Neptune. If there is a detectable difference between Pluto and Neptune, this simple trick shows it immediately. Using a relative measurement like this, there is no need for complex, careful corrections for instrumental factors. Lampland found by comparing Pluto to Neptune that Pluto was much yellowier than its bluish sister. He attributed the different hue to "a [kind of] different atmosphere." Lampland would never know how right he was.

Knowing Pluto is yellowier than Neptune is a nice comparative, but it does not say what Pluto's *true* color actually is. Is it simply that Pluto *appears* yellowish because it is mirroring sunlight? To answer that question, one must record a color measurement taking one additional factor into account: that Pluto's apparent color (like that of every other planet in the solar system) is influenced heavily by the yellow color of the Sun – which is the light source illuminating it. Just as a red lamp makes a room appear reddish, so too the Sun's yellow illumination affects the way Pluto and the other planets appear to us on Earth. So, to determine Pluto's true color, it would be necessary to measure both Pluto's color and the Sun's using the same observational setup.

How does this work? Consider these examples in an imaginary experiment using a pair of red and blue filters. If the Sun is twice as bright in the red filter image as in the blue filter image, but Pluto is four times as bright in the red filter image as the blue filter image, then clearly Pluto is redder than the Sun. Likewise, if Pluto were only twice as bright in the red filter as the blue filter image, then Pluto would be reflecting the sunlight without changing its color, so astronomers would say it reflects all colors equally and is thus gray. And, if Pluto is only a little brighter in the red filter than the blue filter, while the Sun is twice bright in the red filter image as in the blue, then Pluto itself must be rather bluish, so that it can tone down the red color of the Sun when it reflects sunlight back to Earth.

In 1933 Vesto Slipher reported on exactly this kind of measurement. Just three years after Lampland's comparison to Neptune, Slipher had determined something significantly more useful: "Pluto's light in the yellow and red is apparently somewhat stronger than normal sunlight, and so he [*sic*] is a yellowish planet, possibly standing between Mercury and Mars in color."

Over the years, astronomers recognized that color measurements could be inter-compared much more easily if everyone used the same kinds of filters. So, a standard set of five filters was developed to measure the brightness of stars and planets in wavelengths ranging across the spectrum from just beyond the bluest wavelengths the eye can detect to a little bit beyond the reddest. These five filters are known as U (ultraviolet), B (blue), V (visual, actually green), R (red), and I (near-infrared). Comparing the brightness of an astronomical object through any two of these filters gives a quantitative measurement of its color, known as a "color index." By the early 1960s good UBVRI measurements of Pluto and many other objects in the solar system had been made by people like Bob Hardie, and it was possible to say rather precise things, like, "Pluto's B-V index is near 0.85." What does this jargon mean? It simply means that Pluto is just slightly redder than the Sun, and would appear to the eye only faintly more red than a full Moon, which means hardly at all!

Now that we have seen how color measurements are made, it is worth asking what colors tell us about Pluto's fundamental nature, or about its surface composition. As it turns out, not much. Colors give a feel for how a planet may appear, but they just do not provide enough specifics to deduce hard information on the composition of the object's atmosphere or surface. After all, objects as different as our sky and a blueberry are blue, and despite their vastly different nature and composition, both bananas and the Sun are yellow. Likewise, Mars is red, as are roses, blood, and many kinds of beers, but these objects do not bear much compositional similarity to one another. As the saying goes, "you can't judge a book by its cover."

To learn something definitive about compositions, much, much more detailed information than just color is needed. What is required is a spectrum, which is simply a graph of the brightness of an object across a range of finely divided colors, or wavelengths. The idea is this: measure the brightness of an object at 100 or 1000 colors between red and blue (or better yet, infrared and ultraviolet) wavelengths and you will have something into which you can sink your teeth. Why does this work? Because every type of atom, every molecule, every substance imprints its own spectral signature on the light that it emits or reflects. These signatures come in the form of little dips called absorptions, which occur at various specific wavelengths (i.e., colors) unique to each molecular constituent. In effect, the pattern of absorptions for each constituent forms its fingerprint. If a given molecule is sufficiently prevalent on the surface or in the atmosphere of a planet, then it can make deep enough absorptions to be seen in a spectrum.

So, if you want to know what kinds of material lie on Pluto (or Titan, or Triton, or Timbuktu for that matter), just compare the spectrum to those of simple materials, and see how much of each is required to obtain a good match. The technique is called spectral matching, and it is so powerful that whole libraries of templates for pure materials,

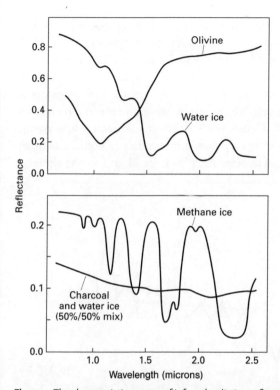

Fig. 2.3: The characteristic spectra of infrared radiation reflected by different materials is an important diagnostic tool for identifying the surface composition of planets and their satellites. These examples show the very different reflectance spectra of olivine (a common silicate mineral), water ice, methane ice, and a mixture of water ice and charcoal (50:50 by weight), and they demonstrate the basics of how the infrared spectra of common materials can be used to identify surface compositions across the solar system. (Note the expanded vertical scale of the bottom panel.)

called "reference spectra," have been measured and then carefully catalogued on computers to make the job easier (Figure 2.3).

Unfortunately, the technology for making spectroscopic measurements of objects as faint as Pluto did not come along until the mid-1970s. One reason for this was that, in order to record a spectrum, the light from the target body must be divided up among many closely spaced colors (usually called channels, or bins). Consequently, it takes a lot longer to obtain a spectrum than to obtain a color image through a filter letting through a broad range of colors, simply because there is less light in each spectral bin once it is been divided up. Worse, to avoid the nonlinearity problems associated with photography, the only option in the 1960s and 1970s was to use PMT detectors. However, as we mentioned above, PMTs have only one pixel. This means you can only use a PMT to measure the light from one spectral color (one bin) at a time. As a result, it is necessary to measure each wavelength in turn until the entire spectral region under study is covered. Hence, most astronomical spectrometers of the 1950s to the 1970s were very slow at measuring an entire spectrum. It could take all night or several nights to build up a strong signal.

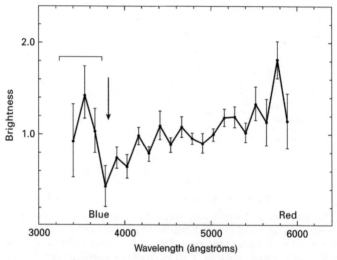

Fig. 2.4: The first spectrum of Pluto, obtained in 1970 by John Fix and colleagues. They measured the brightness of Pluto at 21 points across the visible spectrum. The vertical bars through the points are "error bars," indicating the extent to which each measured point is uncertain. The general trend shows that Pluto reflects more red light than blue light. The three points on the far left (bracketed) turned out to be mistakenly high, making it appear as if there is an absorption dip in Pluto's spectrum (arrowed). (Adapted from J. Fix et al. 1970, *Astronomical Journal* 75)

The first published spectrum of Pluto was reported in 1970 by a small group of University of Iowa researchers led by John Fix, who used a 24-inch telescope in Hills, Iowa. Fix and company's telescope was too small to allow them to get a good result if they broke the light up into too many narrow slices of the spectrum, and all they could manage was a 21-point spectrum from blue to red wavelengths. Limited as that seems now, it was a real step forward.

The spectrum that Fix and colleagues obtained is shown in Figure 2.4. Notice how this spectrum shows that Pluto is brighter at the red end of its spectrum than at the blue end. This is due to Pluto's reddish color, as had been discovered by Hardie's UBVRI photometry.

The second thing to notice about Fix's spectrum is that the three leftmost (i.e., bluest) channels bump up compared to their redder neighbors to the right. One zealous astronomer interpreted the saw-tooth-like shape of the dip between the red slope on the right and the three leftmost channels as an absorption feature due to either iron-bearing minerals or, possibly, silicate rocks and happily went off to write a scientific paper claiming Pluto's surface had copious iron or silicates in it. Nice try, but he had over-interpreted the data.

Simply put, the dip attributed to an absorption feature was not real at all. It was caused in large part by the fact that the spectrometer Fix used was not as sensitive to very blue light and very red light, as it was to light of intermediate colors. As a result, the error estimates (those little bars on each measurement point) in his data were larger at the edge of the spectrum than in the middle. In fact, the apparent increase in Pluto's reflectivity at the three bluest data points is just an artifact of this particular instrument, which tended to show just such an upturn at the bluest wavelengths it could record when measuring faint signals. If you want to get it right, you have to be careful, and that means knowing the detailed ins and outs of how the measuring tools you are using really work. The fellow who leapt to conclusions based on meager and noisy data was a little too quick for his own good. Science is competitive, sometimes leading to premature re-sults that are later proven wrong; fortunately, the competition and cross-checking inherent in the scientific method eventually weed out incorrect results obtained in haste.

Better Pluto spectra in the visible region, with both greater preci-sion and more densely spaced spectral channels, became available as the 1970s progressed. However, the real prize was waiting in the infrared region of the spectrum, which has wavelengths longer than the reddest light our eyes can detect, and is far richer in spectral fea-tures due to solid materials than is the visible band. As a result, most compositional studies of planetary surfaces now employ infrared in-struments.

By the early 1970s infrared detectors were advancing rapidly in sensitivity, largely as a spin-off from Cold War military investments in detectors for heat-seeking missiles. As infrared devices improved, astronomers applied them to successively fainter targets: first to the large satellites of Jupiter, and then to the main satellites of Saturn. These devices revealed that most of these little worlds were encased in shells of water ice. But there were also a few oddballs, like Jupiter's Io, which was practically barren of water and seemed to be littered in sulfur compounds. Equally notable was Saturn's satellite Titan, which is about the size of Mercury, and whose near-infrared spectrum displayed evidence for a methane (CH_4) atmosphere.

Pluto's Own Signature

With the frontier rolling back in leaps and bounds as infrared detector technology kept improving, the desire to obtain an infrared spectrum of Pluto was keen. At the rate that detector technology improvements were becoming available through military development programs in the late 1970s, it looked like the feat might just be achievable by the end of the decade.

However, those astronomers waiting for the better infrared detectors that would be in use at the end of the 1970s never had a chance, because they were scooped by three young astronomers with a clever idea.

These three clever astronomers – Dale Cruikshank, Carl Pilcher, and David Morrison – realized that there was another way to make some inferences about the surface composition using a large telescope feeding an infrared-sensitive PMT with specially selected wideband infrared filters in front of it. Although the information content was crude, Cruikshank and colleagues knew that with wide-band filters, they could get enough signal through to obtain useful results right away, without waiting for a sensitive enough infrared spectrograph to become available.

Their technique depended upon choosing filters in a very clever way to yield diagnostic results. Cruikshank and colleagues recognized that the most common frosts expected in the outer solar system, namely water ice, methane ice, and ammonia ice, reveal themselves by strong absorption features between 1.4 and 1.9 microns. (A micron, equal to 10 000 angstroms or 1/10 000 of a centimeter, is the standard unit of wavelength in infrared astronomy.) Importantly,

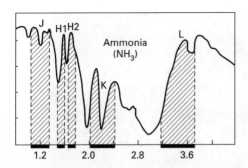

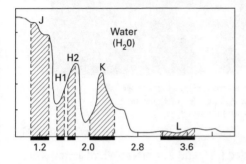

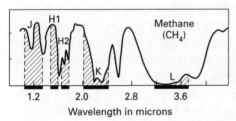

Wavelength in microns

Fig. 2.5: The first evidence for methane frost on Pluto came in 1976 from measuring its brightness in infrared light through a set of specially selected filters. These diagrams show the "windows" corresponding to five of the six filters used by Dale Cruikshank and colleagues, superimposed on the spectra of infrared radiation reflected from three chemically different ices that might have been expected: ammonia, water, and methane. The relative intensity expected through each of the filters follows a different pattern for each ice. By determining which ice fitted their measurements best, Cruikshank and colleagues deduced that methane frost was present on Pluto. (Adapted from Cruikshank et al. 1976, *Science* 194)

each of these ices has its own unique set of absorptions, as shown in Figure 2.5.

Cruikshank (Figure 2.6) and colleagues selected five filters that, as a set, could distinguish between the most likely ices expected on

Fig. 2.6: Dale Cruikshank, who with colleagues Carl Pilcher and David Morrison found the first evidence for methane frost on Pluto's surface. Photo ca. 1996. (J. Mitton)

Pluto. Three of them, called J, H, and K, were standard astronomical filters which extend the UBVRI system into the infrared part of the spectrum. They also constructed two customized filters, which they called H1 and H2. The pattern of brightness measured through this set of five filters could distinguish the differences between water ice, methane ice, ammonia ice, and common rocks.

After requesting and receiving observing time on the large, 4-meter (158-inch) Mayall telescope at Kitt Peak, Cruikshank, Pilcher, and Morrison undertook four nights of intensive Pluto studies in March of 1976. On each night they repeatedly measured Pluto's faint reflectivity through their five filters. After carefully calibrating their data against standard stars, selected for their constant and well-known infrared brightness, they found that Pluto's surface showed no evidence for rock, water ice, or ammonia ice, but strong evidence for something far more exotic – methane ice!

Cruikshank and colleagues' simple and elegant experiment beat the first spectroscopic detection of methane on Pluto by over two years, and was the scientific equivalent of the mouse that roared. For the first time, there was definitive information about the composition of Pluto's surface. And the three young scientists milked the data for all they were worth.

The fact that an exotic ice like methane (frozen natural gas, no less) covers Pluto's surface, rather than water ice, or ammonia, or rock, indicated immediately to researchers that Pluto's surface chemistry was very different from almost all of the distant outer-solar-system satellites that had been spectroscopically explored. Further, the fact that the methane signal that Cruikshank and colleagues saw in their data did not quite match the pure methane signature they obtained in a laboratory experiment using the very same filters, indicated that something else was probably mixed with the methane on Pluto's surface. Further still, Cruikshank, Pilcher, and Morrison pointed out that the existence of methane, which cannot condense to an ice on planetary surfaces except in the cold outer solar system, argued against speculation that Pluto was an errant inner planet or asteroid that somehow found its way to the great deep beyond Neptune.

This basket of findings echoed around the worldwide planetary community and caused a new wave of researchers to sit up and take notice of Pluto. The little planet was developing a reputation for scientific intrigue far beyond its diminutive size.

As 1976 turned 1977, and then 1978, the long-awaited, new generation of infrared spectrometers became available, allowing a far more detailed reconnaissance of Pluto's infrared spectrum than Cruikshank, Pilcher, and Morrison had achieved, but nonetheless, they had been first. In short order, no less than three teams confirmed the methane signature, and opened the door to more detailed analysis of it. At the same time, however, another breakthrough was being made, and it was one that no one had expected.

A Rose, By Any Other Name

By the time the late 1970s arrived, the first really useful fragments of the Pluto puzzle were falling into place, and they indicated that Pluto is not at all the giant world that the other outer planets are.

The evidence that Pluto is far smaller than the other outer planets came from several quarters. The first clue came as the suspected differences between Neptune's predicted position and its measured positions evaporated. With better measurement technology, smaller and smaller estimates of Pluto's mass were required to explain them.

Then, no Plutonian disk could be resolved. Although Earl Slipher's quaint demonstration showed that a small enough planet would show no disk from a great distance, it was worrisome to many that, over the succeeding 45 years, no one had ever, even during rare moments of exceptional atmospheric clarity and stability, seen Pluto as anything but a pinpoint on the sky. And this was not for a lack of trying. Astronomical giants like Gerard Kuiper and Milton Humason tried, again and again. Kuiper even had a special machine called a "disk-meter" built to project tiny little calibration disks next to Pluto's image in the telescopes for comparative purposes. Using the disk-meter on the 200-inch Hale telescope, the largest astronomical telescope the world could then offer, Kuiper concluded that Pluto's diameter was 5900 kilometers (3700 miles). This meant that Pluto was indeed smaller than the giant planets, smaller even than Earth. But as better measurements later proved, Kuiper and Humason's 5900-kilometer diameter was still a severe overestimate.

Other tricks to get at Pluto's true size were tried as well, as astronomers of the mid-twentieth century tried to beat the limitations that their technological bondage imposed. For example, could Pluto's effect on Halley's Comet be measured? If so, its mass and therefore its size could be constrained. Might tiny Pluto someday pass in front of a star? If so, then its size could be directly measured by timing the tiny eclipse. Such ideas deserve an "A" for ingenuity, but yielded little in the way of useful results in their day. It was frustrating, like waiting by a phone that never rings and, by its continued silence, reminds one that the caller is absent.

The first real advance of the modern era came again from Cruikshank's spectroscopy team, who noted that laboratory samples of methane frost appear to the eye to be as reflective as fresh water snow, indicating that Pluto's reflectivity must be very high – so high in fact that it seemed likely Pluto would be still smaller than then expected. Based on the high reflectivity of methane, Cruikshank, Pilcher, and Morrison concluded that Pluto's diameter was between perhaps 1600 and 5300 kilometers (1000 and 3300 miles) – smaller than the Earth's Moon!

Shortly after this result was published in *Science*, space physicists Alex Dessler and Chris Russell parodied the ever-declining mass estimates for Pluto as a function of time. Their conclusion: Pluto was disappearing, and by the early 1980s it would be gone altogether (Fig-

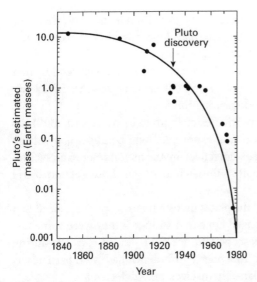

Fig. 2.7: Before Pluto was found, and over the first five decades following its discovery, succeeding estimates of its mass became less and less. In 1980 Alex Dessler and Chris Russell jokingly insisted that a continuation of this trend would lead to the planet's imminent disappearance! It was subsequently determined, as a direct result of the discovery of Charon, that Pluto has only two thousandths the mass of Earth and is roughly 2350 kilometers across. (Adapted from A. Dessler and C. Russell 1980, *Eos* 61)

ure 2.7). This lack of respect, albeit tongue-in-cheek, was about to evaporate altogether.

Fellow Traveler

June 22, 1978 was one of those hot summer Thursdays in Washington, DC, and US Naval Observatory (USNO) astronomer James Christy was preparing to move his family from their apartment to a new home within a few days. Looking for a "short and routine" task to round out his week, Christy had no idea that he was about to make one of the most significant marks on the history of planetary astronomy in the 1970s.

Christy's boss, Bob Harrington, suggested that Christy "measure" a set of Pluto plates as data to feed into a better determination of Pluto's exact orbit. Because he worked for the Naval Observatory, which among its many navigation-related duties is responsible for precise planet tracking, Christy had done this kind of work many times in the past. The Pluto plates Harrington gave Christy to measure[10] had been obtained in Flagstaff at Christy's own request that April and May, in support of a Naval Observatory program of careful, routine planetary orbit determinations.

In total, there were 18 images of Pluto to be measured, and Harrington remarked that it might not take long because many of the plates had been marked "defective" by the astronomers and technicians at USNO's Flagstaff station, where they had been obtained and developed.

As Christy inspected the plates using a microscope, he noticed that the images of Pluto indeed appeared strange – they were somehow elongated, as if the telescope had not quite guided correctly during the exposures, or the atmospheric turbulence had been particularly poor, or perhaps the plates themselves were defective.

Then, Christy realized something. At first it was subliminal, but then it crystallized as a conscious thought: Pluto looked elongated, but the stars on the plates looked perfectly round! The plates were not "defective" – Pluto was.

On the plates taken on April 13, the extension of Pluto's image was to the south; on the plates taken a month later on May 12, Pluto's extension was to the north. Maybe, Christy thought this was some feature locked with Pluto's rotation. Maybe he had discovered a giant mountain on Pluto, but when he estimated the height it would have to be to be visible to him the answer was many thousands of kilometers. Ridiculous, he concluded. No way. He dismissed the idea essentially on the spot.

On second thought, however, maybe a giant Krakatoa was erupting on Pluto? Now that would be something no one had ever seen – a volcano going off on a planet. No, volcanic eruptions like that do not usually last a month. Month. Month? Moonth! "What?" Christy later wrote recalling his thoughts of the moment, "Pluto has a moon?"

Could this be? Was this real? Like Tombaugh 48 years before him, Christy felt the need to have another pair of eyes see the thing. So within minutes, he had corralled astronomer Jerry Josties into the darkened room. Josties agreed something interesting was definitely on these "defective" plates (Figure 2.8).

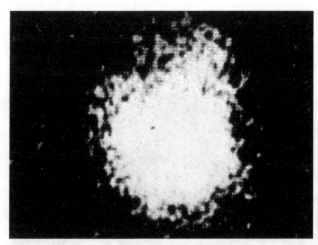

Fig. 2.8: The original 1978 photograph on which Charon was discovered by James Christy. It was taken with the US Naval Observatory's 1.55-meter diameter telescope in Flagstaff, Arizona, just 4 miles from where Pluto was discovered some 48 years earlier. On this image, Pluto is the large white blob and Charon is the small extension to the upper right of Pluto. Although the two are actually separated by over 17 000 kilometers, blurring by the Earth's atmosphere makes the two appear to be a single pear-shaped blob. (US Naval Observatory)

Christy and Josties tried to eliminate the possibility of a moon by brainstorming what else could cause a distant world to look like this. After all, Pluto's image was less than a millimeter across and the elongation was less than half that. It would not take much to create the tiny extension Christy had detected.

Ah, some dirt on the plate at just wrong spot. But this was very rare; how could specks of dirt be on the Pluto images in all the plates, but not on any of the stellar images? Not a plate flaw... Well, maybe it was a background star just off Pluto's side. Christy went to the Palomar Sky Atlas and checked, but found that there were no sufficiently bright stars close enough to Pluto's position in April and May of 1978 to blend with the Pluto images and create the elongation. So it was not a background star unluckily nestled behind Pluto.

Something *was* up, and Jim Christy decided it was time to find his boss (Figure 2.9). Christy remembers that Harrington responded to the claim that the plates might be revealing a satellite of Pluto with one short sentence: "Jim, you're crazy."

The next morning, Friday, Christy still could not let go of the idea. So he went down to the Naval Observatory's collection of archival

Fig. 2.9: James Christy (seated), who discovered Charon, with his colleague Robert Harrington. Harrington doubted the discovery at first but later became convinced Charon was real and used Charon's orbit to determine the mass of the Pluto–Charon system. (US Naval Observatory)

plates, which contained images of Pluto stretching back for decades. He found a set made in June of 1970, and sure enough, the little elongation was not only there, it cycled around Pluto over a span of a week!

This was so close to Pluto's known 6.387-day rotation period that Christy figured it could not be a coincidence. So Christy assumed the orbital period was precisely 6.387 days, the same as Pluto's rotation period, and asked Harrington to calculate on which side of Pluto the little elongation should lie in each of the plates made in June of 1970, some eight years and almost 500 rotation periods before. Nature would have to be perverse indeed to create a fluke of this magnitude. If this checked, the satellite was very probably real.

According to MIT astronomer Rick Binzel, who was then working as summer student at the Naval Observatory, Harrington and Christy worked independently to make this check, so that neither knew what answer the other found. When both had completed their work, they met in Harrington's office. "What did you get?" "Well, what did *you* get?"

As it turned out, Harrington's calculations of where the satellite should be agreed amazingly with Christy's measurements of the elongations seen on the 1970 plates. Less than two hours later Christy found and checked two additional plates made back in 1965. Again the elongation appeared right where Harrington predicted it should be.

The satellite was real, and it was apparently in an orbit that was synchronized to Pluto's rotation period. Barely over a day had flashed by since Harrington had first set Christy to routinely measure these possibly "defective" plates. But in that short span of time Christy had discovered that Pluto has a moon.

While other astronomers at the Naval Observatory in Washington, DC, continued testing the satellite idea against various other alternative explanations, the USNO's Flagstaff station began a program to obtain more observations of the newly found satellite.

Pluto's moon had been found *entirely by accident*. How could this be? Many times in the past astronomers had searched for satellites of Pluto. Indeed, it was such an obvious thing to do that even Slipher, Lampland, and Tombaugh had searched for satellites back in March of 1930, the month they announced the discovery of Pluto. Later, in 1950, Humason (a solid observer of legendary repute, but the same man who missed Pluto itself in 1919) conducted a dedicated search for satellites of Pluto, but without success. Others tried as well. Otto Franz, a talented astronomer who took the 1965 plates that Christy used to confirm the discovery, later said that he had noticed the elongated image of Pluto in 1965, but dismissed it as a result of atmospheric turbulence. Franz had not noticed what Christy had – that only Pluto was elongated; owing to that oversight Franz was still gently kicking himself for decades after Christy's discovery.

Fig. 2.10: A modern ground-based image of Pluto and Charon. Though the two bodies are well resolved, no surface detail can be seen. The presence of Earth's atmosphere limits the capability of all telescopes on the surface of our planet. (Nordic Optical Telescope)

With hindsight, it is clear that Pluto's satellite (see Figures 2.10 and 2.11) was not such an easy thing to find. For one thing, Charon is ten times fainter than Pluto, and it never strays farther than 1/4000 of a degree from Pluto as seen from Earth. Thus, the little companion is usually lost in the glare of Pluto. Further, as astronomer Bob Marcialis later described in an analysis of why Pluto's satellite had not been found previously, several factors had conspired to keep the

Fig. 2.11: A Hubble Space Telescope (HST) image of Pluto and Charon taken in early 1994. This image clearly resolves the two objects as separate disks but does not show surface detail. Later HST images revealed surface markings. (R. Albrecht, ESA/ESTEC, and NASA)

companion a secret for the 48 years since Pluto's discovery. Among these had been the use of small telescopes in early satellite searches. In addition, searchers had taken long exposures to detect faint objects relatively far from Pluto, which oversaturated the Pluto image and wiped out any chance of detecting a close-in satellite. Marcialis also noted that the fact that *any* putative moon was harder to see between the 1930s and 1960s simply because of Pluto's greater distance from the Sun and Earth.

Once Harrington and Christy and other astronomers at the Naval Observatory were convinced that the satellite was real, it fell to Christy to name his find. Within a day of his discovery he offered his wife, Charlene, the honor: "I could name it after you – it could be Charon." He was thinking it rhymed with proton and neutron. What a romantic astronomer!

However, Christy realized within a few days that although it was his to propose a name, the International Astronomical Union's (IAU) nomenclature bylaws stated that he would have to chose from Greek or Roman mythology, rather than name it after his wife. Someone suggested Persephone, the wife of Pluto. Christy liked that, so he went to a dictionary to get himself out of the trouble this could cause with his wife. The search was on for Ch- names in the mythology. To his amazement, there, in black and white on the page, was the name of a different ancient deity: "Charon."

This was too good to be true! In Greek mythology, Charon was the repugnant boatman who rowed dead souls over the river Styx into Hades, where the god of the underworld, Pluto, ruled. Christy realized this was the solution to his dilemma with both the IAU and his wife. Although correctly pronounced "Khar-on," Christy pronounced it "Shar-on," like *Char*lene who promptly got her moon after all.[11]

And so it was done. Pluto's satellite, which had for billions of years remained nameless in its anonymity, had acquired a name less than a week after its discovery by one very careful astronomer.

Harvest: 1978

Charon's discovery was announced on July 7, 1978. If Dale Cruikshank had metaphorically fired the first Pluto shot heard round the world with the discovery of methane on Pluto's surface, then Jim

Christy's discovery of Charon was the second volley. That little foot-note of a planet was getting more and more interesting all the time.

Within a few weeks, observations to refine Charon's orbit were coming in from observatories around the world. The basic orbit was determined to be a simple circle, without any noticeable ellipticity, somewhere between 17 000 and 20 000 kilometers (10 000 and 13 000 miles) over Pluto's surface. Because orbital mechanics dictates that close-in satellites must orbit over a planet's equator, Charon's north–south orbit on the sky meant that Pluto's equatorial plane also pointed north–south, confirming Anderson and Fix's finding that Pluto's polar axis was tipped by at least 50 degrees. In fact, using Charon's orbit as a plumb line, Pluto's obliquity was determined to be about 120 degrees, even more than Uranus's. The planet was not only on its side: something had made it tip 30 degrees farther still.

Using only the 6.387-day orbit period and the roughly 19 000-kilometer (12 000-mile) radius of Charon's orbit, Kepler's third law allowed astronomers to calculate the combined mass of Pluto and Charon. Skipping the arithmetic and cutting to the chase, Pluto's mass was estimated within a few days of Charon's discovery, and it was even tinier than had been expected – a paltry two thousandths of the Earth's mass. That in turn provided a very good estimate of Pluto's diameter: between 1900 and 3200 kilometers (1200 and 2000 miles).[12]

Christy's discovery also halted the decades-long "fall" of Pluto's mass toward zero, but the catch was very close. Compared to both the predicted mass of Pluto and to *every other* planet, Pluto was, well, a miniature.

If a mass of 2/1000 that of the Earth does not quite grab you, then consider this: planet Pluto's mass is less than one-tenth that of our own moon, Luna. Beside its neighboring planets, Pluto is not just a dwarf among giants: it is a flea.

Based on Charon's brightness, which was about 1/6 that of Pluto, it was immediately estimated that, if the pair had similar reflectivity and internal density (a reasonable first assumption), then Charon's mass had to be in the neighborhood of 1/10 of Pluto's, and the satellite was likely to be between 1/3 and 2/3 the size of Pluto.

Nothing else like this pair had ever been seen in the solar system. The diameters of most satellites are smaller than 1% of the diameter of their parent planet. Even Luna in the Earth–Moon system, the previous record-holder for comparable size to its parent, is just a quarter the diameter of the Earth. Charon, however, was 50% of the diameter of Pluto.

Put another way, whereas roughly 50 Lunas could fit within the Earth, only about 8 Charons could fit within Pluto. This in turn meant that the balance point, or barycenter of the planet and its moon, lies

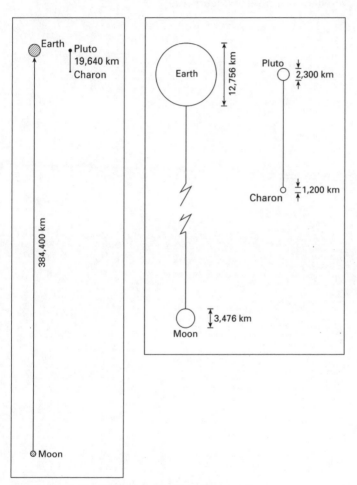

Fig. 2.12: Left: The Earth–Moon and Pluto–Charon systems to the same scale. Charon is a barely perceptible dot. Right: With the planets and their moons scaled up by a factor of six; at this scale, the Moon should be about a meter off the edge of the page.

between the two in open space, which is unique in the solar system. In every *other* case, the planet is so massive compared to its satellites that the system barycenter lies deep inside the planet itself.

Thus, unlike all the primary planet–puny satellite pairs in the solar system, Pluto and Charon turn out to be almost twinned in size, with the satellite half the size of the planet! When this was realized, Pluto–Charon became known as the first true binary planet system discovered in the universe (Figure 2.12).

3
A Distant Dance

"Then I felt like some watcher of the skies when a new planet
swims into his ken." – Keats

Christy and Harrington's discovery of Charon generated new interest
in the study of Pluto. It was not just that the discovery was wholly
unexpected, nor was it the simple, freshness of the find, for satellites
of planets had been discovered scores of times before. It was something different. It was that the long-abstract concept of Pluto – that
far-away pinpoint of light – gained a new kind of reality. With Charon
by its side, Pluto was no longer just the featureless dot it had been for
so many decades. Those little, elongated images of the Pluto–Charon
system gave Pluto dimension, and with that a reality it had previously
lacked. Thanks to Charon, Pluto left the long list of faint embers
drifting far against the Deep, and joined the other planets as places
that the eye remembered as a destination, with extent, and shape,
and just, perhaps, a face.

The Gift of a Swede

Everyone has a favorite planet, and for Leif Andersson it was Pluto.
Andersson was a Swede of little means, who had so wanted to attend
college that he financed it by entering and winning Swedish TV game
shows. After undergraduate school in Sweden, he went to Indiana
University in the USA to study astronomy for a PhD. There he became
fascinated with Pluto, though no one we spoke to quite remembers
why. His PhD advisor, Martin Burkhead, recalls Andersson back
in the early 1970s as "an incredibly talented student, scientifically
independent, and likely to become one of the leaders of his generation
of planetary scientists."

Andersson chose to base his doctoral dissertation on accurate photometry of the icy satellites of the outer planets. Despite the fact that
Pluto was not an icy satellite, Andersson included it on his list of

Pluto and Charon, S. Alan Stern and Jacqueline Mitton
Copyright © 2005 WILEY-VCH Verlag GmbH & Co. KGaA, Weinheim
ISBN: 3-527-40556-9

targets. How could he not? Somehow, in those early days before the depth of Pluto's mysteries was even clear, the Swede had taken little planet Pluto to his heart. Andersson had become a Plutophile, long before being a Plutophile became cool.

Andersson finished his PhD research in 1975, having refined knowledge about Pluto's photometric properties, and took a position as a postdoctoral researcher at the Lunar and Planetary Laboratory in Tucson, Arizona. There, just weeks after Charon was discovered in 1978, Andersson was the first to realize that Charon was about to become the key that would unlock the door that had for decades guarded so many of Pluto's secrets.

We do not know exactly how or where Andersson got the idea, but in the summer of 1978, a revelation washed over him: sometime soon, Pluto and Charon were going to undergo a set of mutual occultations (i.e., events in which they would be moving in front of, or blocking, one another repeatedly) as seen from Earth's vantage point. If Andersson's calculations were right, these "mutual events" would yield precise information on the sizes, densities, and compositions of Pluto and Charon, and even information about their surface markings. The information contained in the lightcurves of those mutual events would be manna from heaven for planetary astronomers.

The crux of Andersson's realization was elegant in its simplicity. As Pluto orbits the Sun, the direction of its polar axis remains almost precisely fixed in space (just as the Earth's polar axis stays pointed in space toward the North Star, Polaris). This "axial stability," which is common to all of the planets, is generated by the enormous rotational momentum inherent in any large, spinning object.

Pluto's pole, for example, has been aimed toward a point between the bright stars Spica and Antares, since long before recorded human history. This stability creates the interesting consequence that we here on Earth, deep down toward the Sun near one focus of Pluto's orbit, see Pluto's pose (which astronomers call its "aspect") from different angles as it moves around the Sun. Because Pluto *is* tipped over on its side (something John Fix and his team had discovered early in the 1970s), it displays to us a dramatic range of aspects during each 248-year orbit (Figure 3.1).

What Andersson realized before anyone else is that, as Pluto's aspect changes, Charon's orbit (which is locked into Pluto's equatorial plane, and therefore changes aspect with Pluto) is also seen from

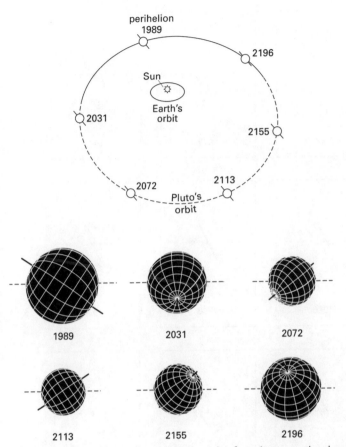

Fig. 3.1: Pluto's rotation axis is tilted at an angle of 122 degrees to the plane of its orbit, making Pluto appear to lie on its side. As with any planet, Pluto's spin axis stays pointing to the same direction in space (top) as it orbits the Sun. However, as Pluto progresses around its orbit, the view of Pluto from Earth – its aspect – gradually changes (bottom). These schematic diagrams (not to scale) show Pluto's location in its orbit at intervals through the 248 Earth years of a Plutonian year, and the corresponding views of Pluto from Earth. Note also how the apparent size of Pluto varies with its changing distance from Earth during the orbit. The section of Pluto's orbit shown with dashes is the portion that lies south of the plane of Earth's orbit.

various angles (Figure 3.2). Sometimes it will appear to circle Pluto like the ring around a bull's eye and other times, when we see the orbit more edge on, the circle will have collapsed to an ellipse, and then, briefly, to a line when it lies completely edge on as seen from Earth. Note that Charon's orbit does not change, but the angle we see

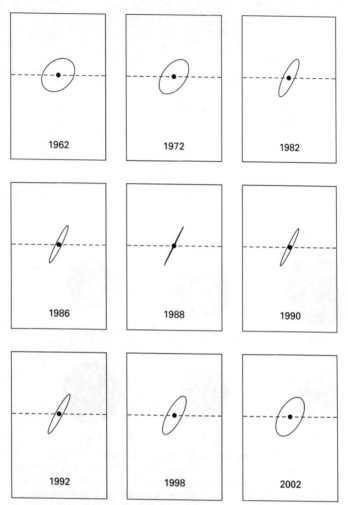

Fig. 3.2: The perspective from which we see Charon's orbit around Pluto gradually changes as Pluto travels around the Sun. This series of diagrams shows the orientation of Charon's orbit as seen from Earth at various dates between 1962 and 2002. The tiny ellipse that Charon's orbit sweeps out on our sky diminishes in size as Pluto's distance from the Sun increases during the course of its orbit, but the effect is not great enough to be detectable in the period illustrated here.

it from does. The key consequence of this, Andersson recognized, is that twice each 248-year Pluto orbit, precisely 124 years apart (half of Pluto's orbit period), Charon and Pluto will undergo a few years when each alternately blocks the other as seen from Earth, creating a series of scientifically valuable occultations.

Andersson estimated that each of the occultations would last up to a few hours during Charon's 6.387-day orbit around Pluto. The earliest of these events would just clip the limb of the other body. Then, over a period of three years, as the alignment perfected itself, the plane of Charon's orbit would cut longer and longer traverses across Pluto, until finally Charon would transit the diameter of Pluto in what astronomers call "central events." In a central event, we on Earth would see the whole of Charon's disk (and its shadow) moving across Pluto's face for a few hours. Likewise, when Pluto's disk moves in front of Charon, Charon would be completely engulfed and hidden for a few hours.

Based on the crude size estimates for Pluto and Charon that were then available, and the altitude of Charon's orbit, Andersson predicted that the prized central events would take place for about two years in the middle of the series of mutual events. After that, the changing relationship between Pluto's position and the Earth and Sun would begin to make less well aligned, and therefore shorter and shorter, non-central events, until about three years after the last central event, the last grazing event would occur, and the far-away eclipse waltz would end. All told, the Pluto–Charon occultation "season," as astronomers called it, would last between five and six years.

From the work of John Fix, Andersson knew roughly where Pluto's pole was pointed, and what Charon's orbit was, and from that he calculated something truly amazing: *The events were likely going to occur in the 1980s.*

What luck! The event season lasts only five years or so and occurs only twice each 248-year Pluto orbit. Andersson could well have discovered that the mutual occultations between Pluto and Charon had last taken place virtually any time during the six long decades between Pluto's discovery and Charon's, and that we had missed them. He could have found that the mutual events had occurred even further back, while Lowell and Pickering were searching for Pluto, or before that, when the great men were but boys. But instead his numbers indicated that the events would begin *almost immediately.* Sometimes Nature provides a gift, and this was one of those times. By shear luck, Charon had been discovered about 120 years after its last shadow dance with Pluto, and, therefore, on the eve of a new set of such events, pregnant with scientific promise.

Like many pregnancies, those observing the expectant couple did not know exactly when the due date was. Why? Because both the direction of Pluto's polar axis and the inclination of Charon's orbit were each uncertain to a small degree. Neither Andersson nor anyone else could predict exactly which year the events would begin. Andersson's first calculations indicated a start in 1979 or 1980, but orbit and pole direction refinements motivated by the promise of the impending events soon revealed that they probably would not begin until at least 1982, and perhaps not until 1985. Still better predictions were not on the cards, given the uncertainties inherent in the data fed into the calculations, and so it fell to the observers to laboriously hunt the events down, searching year by year until the first one was bagged. Once an event was observed, however, it would be possible to predict accurately future events, since they would occur at 3.194-day intervals – exactly half a Charon orbit apart.

The Dance, Dissected

What would it be like to be there, on Pluto, during one of the mutual events? At the beginning, Charon's 4-degree-wide disk would approach and then begin to move over the bright but tiny, distant Sun. As it did, Charon's shadow would race across Pluto's frozen terrain at roughly Charon's orbital speed, some 220 meters per second (about 730 feet per second). During an event, Pluto's snowy surface would be lit only by starlight and the faint glow of Charon's dark side, itself softly illuminated by yellowy sunlight reflected off the reddish annulus of Pluto that remained out of Charon's 1200-kilometer-wide shadow. After a few minutes or hours of darkness, (depending on the location on Pluto's surface and how deep into the six-year event season it was), the Sun would re-emerge to reassert its diluted light and feeble warmth.

As seen from the warm confines of Earth, the same event would slowly unfold through four phases. It begins with *first contact*, when either the occulter or its shadow (depending on the angle from which the event is being seen from Earth) first impinges on the occultee. *Second contact* then occurs when the occulter and its shadow lie entirely

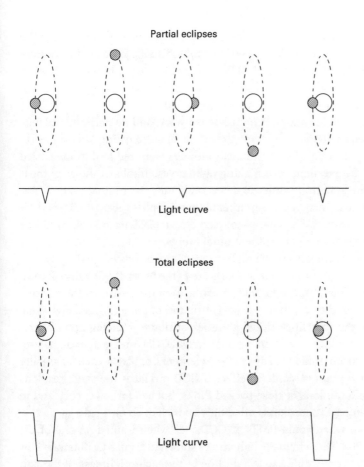

Fig. 3.3: The schematic shown here is a representation of the shape of Pluto and Charon's lightcurve for mutual events involving partial eclipse (top) and total eclipse (bottom).

on the occultee. *Third contact* occurs when the occulter or its shadow begins exiting the occultee. Finally, the event ends with *fourth contact*, when the last slice of the occulter or its shadow leaves the occultee. The so-called period of *centrality* is the interval between second and third contact, when Charon and its shadow lie entirely on Pluto (or, in a Charon occultation, when Charon is entirely hidden by Pluto). Of course, during the first and last years of the mutual event season, when the events were grazing rather than central, only the first and fourth contacts would occur (Figure 3.3).

If the Hubble Space Telescope (HST) had been flying back in the early 1980s, it would have been possible actually to watch the progress of each event as the two bodies seemed to merge, to see the shadow of the sunward body fall on its partner's surface, scoot across the to the opposite limb, and then make its exit.

But HST remained an unfinished work, and no Earth-bound telescope could then overcome the effects of our blurring atmosphere to resolve Pluto. As a result, astronomers were reduced to their tried and true technique of making lightcurves, graphs of the brightness of the system, in order to detect and study the mutual events. Yes, simple lightcurves using either photomultiplier tubes (PMTs) or their successors, called charge-coupled device (CCD) cameras, would be the tool used to probe the mutual events.

CCD cameras are the miniaturized, modern-day digital equivalent of photographic cameras; CCDs are at the heart of every digital camera. They came into widespread astronomical use in the 1980s[13] because they combine the best attributes of the PMT (i.e., digital nature and strict linearity) with the best attribute of photographic plates: they produce a two-dimensional image. CCDs create their digital images from millions of tiny picture elements, or pixels, each far smaller than a grain of sand. In effect, a CCD produces the same result as would millions of close-packed PMTs, but at a far lower cost, and in a compact space that is often only an inch or so on a side.

But what would PMTs or CCDs see when trained on one of the events? The schematic lightcurves shown in Figure 3.3 illustrate how each event would cause a decline in the total brightness of the pair (since part of one body would be blocked from view). The decline would begin at first contact, progressively increase until the event reached its deepest stage, and then reverse itself as the event waned. For the first events, it was predicted that the eclipsed sliver would be so small that the total decline in brightness would be hard to detect, perhaps only one or two percent, and that would have to be detected against the ever-present change in lightcurve amplitude due to Pluto's rotation.

Within months of Leif Andersson's prediction, the power and potential of the onrushing event season became clear, and dozens of astronomers began to organize observing plans. Tragically, however, Andersson was denied the opportunity to take part in the very events he himself had forecast. Andersson contracted and quickly succumbed to lymphoma in 1979, at the age of 35. Had he lived, he would probably have been one of the central characters of the pages that follow. Instead, he remains largely unknown, despite the fact that he launched the greatest Pluto research bonanza of all time.

Searching for Shadows

By the time the calendar rolled into 1982, groups of astronomers in Hawaii, Arizona, California, Texas, Germany, the Soviet Union, and South Africa had planned extensive observational campaigns to detect the onset of the Pluto–Charon mutual events. These searches were not easy undertakings. Not only were predictions of the year and month of the first events uncertain, but uncertainties in Charon's precise orbit meant that the timing of each possible event had its own inherent uncertainty of up to a few hours (equivalent to a couple of percent of a Charon rotation period). As a result, observers not only had to face the fact that they would probably have to observe dozens of candidate events over several years before they might snag their first one, but also that each event could occur within a wide window of time on a given night. Patience with Pluto became a virtue newly appreciated.

Despite the hard work of several groups, no events were detected in 1982, or 1983, or early 1984. Each year there were over 100 potential events, about a third of which would occur during darkness at one of the observatories where patient people searched. But no one observed even the slightest hint or the subtlest signature of the first stages of the on-coming ballet between Pluto and Charon. Thus 1984 too ended without reward, and some astronomers were beginning to wonder whether the events would really take place at all. Prediction calculations were rechecked: indeed the events should now begin very soon, but when? When?

Tally Ho Shadow!

Bonnie Buratti, then a young astronomer with a new job at JPL (Jet Propulsion Laboratory) was the first to spy something, early on the morning of January 16, 1985. Buratti was observing Pluto with astronomer Ed Tedesco at Palomar, high above the Los Angeles basin. Tedesco had long been involved in the search for the first mutual events, and had asked Buratti to help with some of his observing. As it happened, they seemed to stumble onto Charon crossing over Pluto in the early, predawn sky – fully three hours before the nominal time of a predicted event for that day.

What Tedesco and Buratti's data contained was a subtle signature. This had been expected since the first mutual events would only be grazing. But what they observed was *so* subtle: just three lightcurve data points, each depressed only a few percent from what they normally would have been. Was this the real thing, or was it something else?

Tedesco and Buratti had checked the brightness of a pair of comparison stars in between each Pluto measurement and the standard stars had not changed in intensity the way Pluto had. This argued that the signature they saw in their data might indeed be real. However, the fact that the event occurred three hours earlier than expected was worrisome. It was widely known that the exact timing of predicted events could be off by an hour or more, but the estimated probability that an event would occur this far off the mark was less than one in a thousand. Equally troubling was the concern that Buratti and Tedesco's CCD had been running low on coolant just as the shallow, three-point signature occurred, suggesting the possibility that the whole thing was simply an instrumental change, rather than a change in Pluto.

All told, their possible Pluto event was an ambiguous, frustrating kind of thing that begged for corroboration. However, as luck would have it, no one else had set up to observe the January 16 event, so there was no second dataset to check against. Without more data, no one could be sure if the event had been real.

Buratti and Tedesco passed the news of their "event" on to other mutual-event watchers by the already existing scientific e-mail networks of the mid-1980s. Buratti and Tedesco noted their concern about the warming of their CCD and the "event's" early occurrence, but all the same, they offered their data as tentative evidence that a mutual event might have finally been detected.

Murphy's law, of course, always applies in situations like this: as it turned out no one was scheduled to observe another potential event on a large telescope for about a month! Why? As luck would have it, many of the succeeding events over the next few weeks would occur in daylight, and some would occur with Pluto too close in the sky to our (very bright) Moon for observations to be feasible.

So salvation had to wait until February 17, when a young and exceptionally talented University of Texas PhD student named Richard ("Rick") Binzel was scheduled to observe another predicted event. Binzel, using an event time prediction based on the Buratti–Tedesco findings, mounted his PMT on the 0.9-meter (36-inch) diameter telescope located at McDonald Observatory in the Davis Mountains of west Texas, and caught a two-hour mutual event through crisp, clear west Texas skies.

Binzel had started observing Pluto as early as possible that night, and captured an entire event from start to finish. He analyzed his results as they came in, on a programmable hand calculator, no less. The data, showing a decline and then symmetric rise in Pluto's brightness looked convincing. Binzel wanted to be sure, so he rechecked his calculations twice. By 9 a.m. he was convinced he had a confirmed event.

The person Binzel called first was his PhD advisor, the Director of McDonald Observatory, Harlan Smith. From his home in Austin that Sunday morning, Smith asked Binzel, "Are you *sure*?" Binzel said yes, and Smith did not challenge him. Smith trusted Binzel's talent, and his word. Realizing the import of the news that the mutual event season was finally on, Smith replied, "OK, we'll do a press release." Binzel then called Charon's discoverer Jim Christy to tell him the exciting news personally. That done, Binzel began sending e-mail to some of the other observers involved in the search for the first mutual events. Within a day the news was public, wafting through newspapers across the world.

Exactly 3.194 days after Binzel's first event, on February 20, another young astronomer, the University of Hawaii's David Tholen, caught Charon falling into Pluto's shadow using a PMT on the 2.24-meter (88-inch) University of Hawaii telescope on Mauna Kea in Hawaii. That second event, coming right on time based on the anchor point event Binzel had observed, removed any possible final doubts anyone could have: The dance had begun![14]

Within weeks, Binzel, Tholen, Tedesco, Buratti, and their JPL colleague Bob Nelson had teamed together to write up their various findings and submit them to the prestigious scientific journal *Science* (Figure 3.4). Binzel and colleagues reached two important conclusions from the datasets they had gathered from the first mutual events.

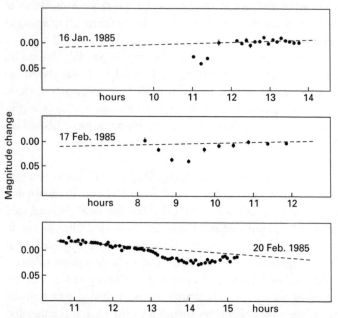

Fig. 3.4: Observations of the first three mutual event lightcurves observed, in 1985. The black dots are the measurements of the change in the combined brightness of Pluto and Charon together, plotted as time sequences during the events. The dashed line shows what was expected had there been no eclipses. The string of dots dips below the dashed line during each event. The time is given in Universal Time (UT) for all three dates. These three lightcurves have been aligned one above another in relation to Charon's position in its orbit. This shows that the dips are linked to the relative position of Pluto and Charon and are therefore unlikely to be caused by anything other than the anticipated mutual events. The January 16, 1985 observation (see main text) was made by Ed Tedesco and Bonnie Buratti with the 1.5-meter reflector at Palomar Observatory; the February 17 event was recorded by Richard Binzel at the McDonald Observatory; the February 20 observation was made by David Tholen on Mauna Kea, Hawaii. (Adapted from Binzel et al. 1985, *Science*, June)

The first scientific result followed from the finding that the sizes of the dips in the January 16 and February 17 event lightcurves, when Charon partially obscured Pluto, were significantly greater than that in Tholen's February 20 lightcurve, when Pluto partially obscured Charon. From this, the five scientists concluded that the terrain on Pluto that had been eclipsed by Charon's shadow must be more reflective than the terrain eclipsed on Charon by Pluto's shadow. This was the first piece of evidence foreshadowing just how strikingly different Pluto and Charon actually are. There would be much more evidence to this effect from later events.

The second result the five researchers obtained came from the fact that the difference in time between Pluto's eclipse of Charon and Charon's eclipse of Pluto was within minutes of the precise, 3.194-day half-orbit period of Charon. The fact that the timing was so exact meant that Charon's orbit was not just roughly circular, as was already known; it was all but exactly circular, perhaps not deviating by more than a few kilometers over its nearly 40 000-kilometer-wide span!

The Pluto–Charon mutual events were yielding quantitative results. Over the next six years, as the events progressed from the shallow graze first detected in 1985 to the central events of 1987 and 1988, and then back, towards the final grazes that occurred in 1991, observers around the world became involved in the careful prediction of new events, and their observation. Others also became involved in the synthesis of data embedded in the lightcurves racing across space from the frozen Plutonian wilderness, five billion kilometers (three billion miles) from Earth. In all, over 20 telescopes and 40 scientists ultimately took part in this work, which yielded a bonanza of new insights into the solar system's distant, dynamic duo. Let us see what they learned.

When Seconds Equal Centimeters: Measuring Pluto's True Size

Astronomers who point their telescopes skyward to study stars and planets have some important things in common with those high-tech spies who point telescopes downward from Earth orbit. In both cases, the observer can look but not touch. So too, both kinds of observer must often resort to wit and clever device to substitute for the limitations of their instruments. For exactly this reason, astronomers have long valued the power of mutual events between binary stars as a tool for leveraging limited telescopic resolution. So when planetary astronomers first learned of Andersson's prediction that Pluto and Charon would occult one another in the 1980s, they salivated as if they were Pavlov's dogs.

As we described in Chapter 2, Pluto's size and great distance conspire to make its apparent size on the sky as seen from Earth truly minuscule – just one-tenth of an arcsecond (a 36 000th of a degree), something like the size of a grain of sand viewed from a distance of about a kilometer (1100 yards). This was far too small for even the best ground-based telescopes of the 1980s to resolve through the turbulence of the Earth's atmosphere.[15]

What a frustration! Fifty years after Pluto's discovery, its diameter had never been directly measured, so its size still was not well determined. In fact, Pluto stood alone among the nine planets in this regard. Every other planet was either close enough, or large enough, or both, to be easy to resolve in a telescope, and therefore had a well-known diameter. But no one knew exactly how big Pluto was.

Indeed, as of 1985, the best available information on Pluto's size came from an indirect source – a calculation. Using the mass determination that resulted from Charon's discovery, it was possible to estimate Pluto's size if one only knew Pluto's density – that is, how much mass was packed in an average cubic centimeter (or inch) of the planet. The fact that Pluto's density was completely unknown meant that one could not constrain its diameter very well. Using the highest and lowest plausible densities imaginable for Pluto, it was calculated that Pluto might be as small as 1900 kilometers (1100 miles) or as large as 4300 kilometers (2700 miles).

Amazingly, back in the early 1980s, Charon's diameter was better known than Pluto's! How's that? For years, astronomers had tried to catch Pluto in the act of occulting (i.e., blocking) the light of a star, so that its size could be determined from the length of time the star was blocked and its rate of travel across the star, both of which are easily measured. Several such events had been predicted to occur but Pluto never actually cut across the various stars astronomers hoped it might. The thing kept missing by a maddening few ten-thousandths of a degree or so. During one such widely anticipated event in April of 1980, South African astronomer Alistair Walker caught Charon occulting the star Pluto should have snagged. This allowed him to estimate Charon's diameter very accurately. Knowing that his photometer showed that Charon had blocked the star for 50 seconds, and the speed of Charon's motion across the star's face, Walker was able to determine that Charon's diameter was almost exactly 1200 kilometers (750 miles).

The mutual events would allow Pluto's size to be determined by turning the difficult (indeed, then insurmountable) problem of resolving Pluto's size in a telescope into a much easier problem – one of timing an occultation between Pluto and Charon. Then, by measuring the interval between first and third contact as Charon glided over the surface of Pluto, and knowing Charon's orbital speed (about 220 meters per second), one could quickly calculate the distance Charon traveled across Pluto. For example, if the time from first contact to third was precisely three hours, then the distance traveled was 2400 kilometers (1500 miles). As easy as that! No wonder so many astronomers competed to observe the events. It is simple enough to almost be a scout's merit badge project.

Well, maybe. It would take a goodly sized telescope and a thorough knowledge of PMTs and data reduction for the scout to do the experiment. And the scout would have to account for some pesky, real-life complications. For example, the early Charon events were just grazes of its shadow across a short segment of Pluto's disk, rather than diametric (i.e., full) transits across its diameter. Therefore, our scout would have to know Charon's orbit accurately enough to tell which part of Pluto was being eclipsed (i.e., in Charon's shadow). Otherwise, it would not be possible to determine Pluto's diameter until all of the events were seen and the longest (across the diameter) was picked out.

Maddeningly, of course, knowing which part of Pluto was being occulted by Charon in turn depends in part on knowing Pluto's size, which is what one is trying to determine. And there were other complications as well, such as shadows that cause the events to appear longer than they really are. That is, because the Sun and Earth as seen from Pluto are offset by up to 2 degrees, the line of sight from Earth can see over Charon's "shoulder" to the place where Charon's shadow falls on Pluto. As a result, the event lightcurve can start earlier or end later than the actual time of first and fourth contact because of shadowing. Better remember to put that in your calculation, and there is more, little scout. What if Pluto (or Charon for that matter) is not perfectly round? And does not the angle we view the events from change slightly during each event, as the Earth moves (at the modest clip of 100 000 kilometers (68 000 miles) per hour)? And what if Pluto has an atmosphere with hazes that might affect the lightcurves? And so on. So, a scout could take the data, if a very good telescope was available, but the scout's parents would probably have to help with data analysis.

We have not run across any reports of scouts trying to measure Pluto's diameter for a merit badge, but we do know that several astronomers did. As it turned out, the single biggest headache (or challenge, depending on your perspective) in determining the diameters of Pluto and Charon from event timing was that the accuracy of the derived diameter depends directly on how well the size (semi-major axis) of Charon's orbit is known. By the mid-1980s Charon's semi-major axis (which is the distance it orbits above Pluto's center) was known to be between 19 000 and 20 000 kilometers (roughly 12 000 and 12 500 miles), with an uncertainty of just 5%. Still, that uncertainty meant that even if everything else that went into the calculation of Pluto's size from a mutual event were known perfectly, Pluto's radius would be uncertain at the 5% level.

In recent years, new data, much of it based on HST results, have allowed Charon's semi-major axis to be determined to within about 1.5% (results range from 19 400 kilometers (12 125 miles) to 19 700 kilometers (12 313 miles)). Combining these numbers with the timing data from mutual events gives a diameter for Pluto somewhere near 2360 kilometers (about 1475 miles), with uncertainties of just a few dozen kilometers (perhaps 40 miles). The cognoscenti are still unable to do better.

Still, the mutual events proved something that was both puzzling and laden with possible promise as a clue to Pluto's origin: Pluto's diameter is far smaller than any other planet's. By comparison to Earth, for example, Pluto would hardly stretch across the central USA. In fact, Pluto is so small that it is even smaller than the Moon, and six other satellites in the solar system: Ganymede, Callisto, Europa, Io, Titan, and Triton. Pluto was strikingly miniature (Figure 3.5).

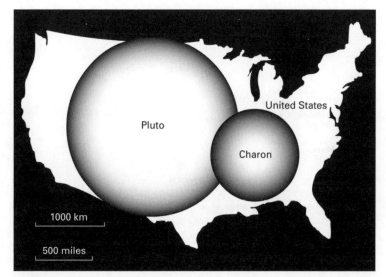

Fig. 3.5: The sizes of Pluto and Charon are shown in relation to the size of the USA. The best estimate of Pluto's apparent diameter is near 2360 kilometers (1475 miles). The best figure for Charon is near 1200 kilometers (750 miles). Both estimates are thought to be accurate to within about 40 kilometers (25 miles).

Bounty

Knowing the sizes of Pluto and Charon opened up the door to other insights about them. One of the first things these quantities provided was the ability to translate the brightnesses of Pluto and Charon into estimates of their surface reflectivity, or albedo. This could not be done before their sizes were known, because the calculation of an albedo depends on how much reflecting area each world presents.

Using the sizes they determined from mutual event data, Dave Tholen and Marc Buie found Pluto's average albedo is about 55% at visible wavelengths. Charon, by comparison, is significantly less reflective, with a visible albedo near 35%.

Pluto's 55% visible light reflectivity is pretty high compared to most of the objects in the solar system. For comparison, the Moon's average reflectivity is about 12%, and most icy satellites have values that hover around that of Charon. In fact, what the 55% albedo translates to is the implication that Pluto's surface is about as reflective as freshly fallen snow. And compared to the freshly driven Colorado ski-powder appearance of Pluto, Charon's surface might be thought of as week-old Boston snow.

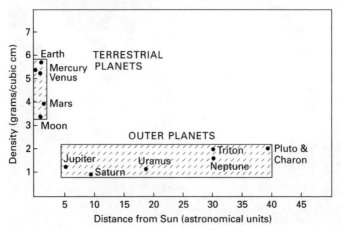

Fig. 3.6: The densities of the nine planets and a few other major objects are shown here in relation to their distances from the Sun. Note that there are two distinct groups: (1) the inner, dense, rocky planets and (2) the outer, less dense, giant planets along with Pluto–Charon and Triton.

Another finding derived directly from the mutual event diameters was a measure of the bulk density (i.e., the average amount of matter packed into each cubic centimeter) of Pluto and Charon. This quantity, which is computed by dividing an object's mass by its volume, is fundamental to making estimates of what a body is likely to be composed of. If, for example, an object's density is high, then it probably contains a good deal of rock. Conversely, if an object's density is low, it is likely to contain a large fraction of lighter materials, like ices, in its interior.

So, knowing the total mass of the Pluto–Charon system by applying Kepler's third law to Charon's orbit (see Chapter 2) and the sizes of Pluto and Charon from mutual event lightcurve timing, it was possible to calculate the average density of Pluto and Charon together.[16] The result: Pluto and Charon have an average density of about 2 grams per cubic centimeter, which is about twice the density of water ice (at 1 gram per cubic centimeter), but significantly less than typical common rocks (at 3 grams per cubic centimeter) (Figure 3.6).

The fact that Pluto and Charon have an average density near 2 grams per cubic centimeter means that the pair cannot be pure ice, but are instead a mixture of ice and rock.

To constrain the interior composition more quantitatively, two teams, one led by NASA's Damon Simonelli and the other led by Washington University's Bill McKinnon fed the average density of the pair into computer models describing the basic structural equations of planetary interiors. From these models, the two teams each determined that the Pluto–Charon pair is probably composed of a mixture consisting of about 65% of light, rocky minerals and about 30% of water ice (the most common ice in the outer solar system), with traces of heavier material (like iron) and lighter material (for example, methane ice, which has a density about half that of water ice).

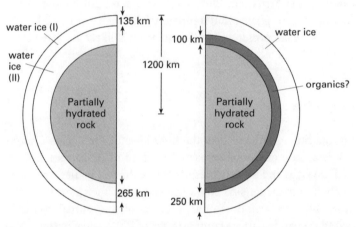

Fig. 3.7: Two possible models for the interior structure of Pluto. In both cases, the surface is covered by a thin layer of frosts, a few kilometers or tens of kilometers thick, which is not shown here. The model on the left is based on an average density on the low side of current estimates (1.735 grams per cubic centimeter) with 54% of the volume being rock and the rest ice. The lower part of the mantle in this case is a form of water ice (ice II) created under high pressure. The model on the right assumes a slightly higher density (1.85 grams per cubic centimeter) and a layer of organic material under the water ice mantle. Owing to the higher density, the proportion of rock in this model is higher, some 70%. (Adapted from W. McKinnon et al. 1997, in *Pluto and Charon*, University of Arizona Press)

McKinnon took this result a little further in a later study with his collaborator Steve Mueller of Southern Methodist University in Dallas. What they wanted to know was whether or not Pluto's interior might have differentiated. Differentiation is the term used when a planet's interior is arranged into a set of layers, with the densest material at the center and the least dense material on the surface, rather than having a more or less uniform mixture of ice mixed with dirt and rock. Diagrams of such objects tend to resemble a kind of onion-skin arrangement. Many planets and some of the large satellites of the outer planets have differentiated into a layered core–mantle–crust structure like the Earth. So, is it likely that Pluto's interior has distinct layers? The best test would be to fly a spacecraft by Pluto to measure the subtle difference in its trajectory that a layered versus a homogeneous Pluto would cause. Lacking that, McKinnon and Mueller ran computer models. What they found was a big "maybe." According to their calculations, Pluto may just barely be large enough and dense enough for its interior to have differentiated (Figure 3.7); or it may not have. As it turns out, Pluto's size and mass put it close to the borderline where planets differentiate under their own weight. To know for sure, according to McKinnon, we will likely just have to go to Pluto and find out.

He and She

So, the little graphs of how the light level of Pluto plus Charon changed as one or the other passed in front of its partner had revealed the sizes and reflectivities of each of the two bodies, and their combined average density. Not a bad start. For their next trick, the eclipse observers would use the very best of the mutual events to reveal the separate surface compositions of Pluto and Charon.

They say that beggars can't be choosers and indeed the shimmering of the Earth's turbulent troposphere had, for centuries, made astronomers beggars when it came to achieving high resolution. If the atmosphere above us blends together the light from Pluto and Charon into some convoluted average of the two, then how could a ground-based observer cleanly split the pair to reveal the light from each individually? By the 1990s there were new technologies finally to beat the atmosphere at its own game, but in the 1980s this was not possible. For the beggars to become choosers, a little ingenuity would be required.

It should not have taken a rocket scientist to realize this, but in fact it did. If you wanted to separate the spectra of Pluto and Charon, all you had to do was wait for a central event in which Charon was completely hidden behind Pluto. During those rare and precious periods, any telescope big enough to gather a good measurement would see Pluto alone, without any "light contamination" from nearby Charon.

Bravo! But better yet, by taking measurements of the combined signal from the pair before or after the event, and then subtracting away the pure Pluto signal obtained when Charon is hidden behind Pluto and therefore out of the picture, one could obtain the difference between "Pluto + Charon" and "Pluto only," thus revealing Charon alone as well.

The race to unravel the separate compositions of Pluto and Charon thus came down to obtaining high-quality visible and infrared spectra between second and third contacts of a central event. The main limitation in accomplishing this was that the Pluto-only spectrum had to be taken during the short, two-hour interval when Charon had disappeared.

If the mutual events between Pluto and Charon had occurred in the 1970s there would not have been much hope for this kind of experiment, because the available detectors just were not sensitive enough to obtain a good spectrum of Pluto in the time available. However, by 1987 when the first central events completely hiding Charon were about to take place, the capabilities of the state-of-the-art instruments sensitive in the compositionally diagnostic infrared region had improved by a factor of 40 to 100 times over those that Cruikshank had at his disposal on Mauna Kea in 1976. With this kind of sensitivity and the onrushing opportunity to separate Pluto's composition from Charon's for the first time, planning for these observations began more than a year in advance. By the time of the first total disappearance of Charon visible at night, in March of 1987, the Plutophiles were out in force.

One observing group was led by graduate student Scott Sawyer at McDonald Observatory, in the Davis mountains 7000 feet above west Texas. The same night, Marc Buie, by then a postdoctoral worker at the Institute for Astronomy in Honolulu, was ready on Mauna Kea, 14 000 feet above the shores of the big island of Hawaii. Meanwhile, Ed Tedesco and his colleague John Africano perched themselves at a telescope atop Kitt Peak, Arizona, to study the event.

But these three groups were not the only ones that set up shop. A careful, two-pronged attack was also planned by University of Arizona graduate student Bob Marcialis, who sat ready with professors George and Marcia Rieke in the control room of the Multiple Mirror Telescope on Mount Hopkins, high above the southern Arizona desert. In addition, Marcialis was collaborating with University of Arizona professor Uwe Fink and his student, Mike DiSanti, who manned a telescope atop Mount Bigelow in central Arizona.

Altogether five groups set up to capture the fleeting four-hour-old photons showering to Earth from the first observable central event. And why not, for the prize was worth the effort.

Both Sawyer's group and the Fink/DiSanti team were operating visible-wavelength spectrometers at the focus of their telescopes. These spectrometers would be used to determine how much of Cruikshank's methane signature was due to Charon. They would accomplish this by monitoring the change in one of methane's strongest spectral absorption bands – near 0.89 microns – throughout the event. Buie on Mauna Kea and Marcialis on Mount Hopkins had each set up infrared PMTs designed to plumb the rich infrared region for what it might have to tell. The event was to begin in the early evening of March 4, and would last several hours.

Twilight fell first on the Texas site, but clouds covered the sky, and a rain briefly fell. Sawyer and his co-workers at McDonald thought they would miss the disappearance of Charon behind Pluto altogether, but thankfully the skies cleared shortly after first contact. All three Arizona teams enjoyed nearly perfect weather that evening, and all three obtained good data. But on Mauna Kea, the winds were so high (120 mph!) that Buie had to close the telescope dome and give up before any data could be taken. Although the skies were clear, he feared that staying open in the wind could damage to the telescope.

It did not take long for the various teams to reduce the highly anticipated data. In fact, the first results were officially announced just one week later at a scientific conference in Tucson, Arizona, when Bob Marcialis (Figure 3.8) took the podium on behalf of both his team and the Fink/DiSanti team. Marcialis is a short and stocky man with deep passions for both softball and science. That day, he swung his research bat three times, making solid hits with each swing.

Marcialis first told the hushed audience that he and his co-workers had found that the methane absorptions first detected by Cruikshank and friends originated entirely on Pluto's surface. This was clear because the strength of the methane absorptions had not weakened

Fig. 3.8: Bob Marcialis ca. 1995. With co-workers he discovered that the signature of methane frost from the Pluto–Charon system is due only to Pluto, while Charon shows evidence for the presence of water ice. Marcialis was also the first to work out a pattern of surface markings on Pluto that would be consistent with its rotational lightcurve. (Photo: Maria Schuchardt)

when Charon disappeared. Marcialis also revealed that the visible-wavelength spectrum of Charon obtained by Fink and DiSanti was both featureless and without any perceptible color or any methane absorptions of its own. Unlike the reddish reflectance of Pluto, the color of the terrain observed on Charon during the event was a pallid gray. Sawyer and his team would later report the same (Figure 3.9). Finally, Marcialis let go his home-run hit: by subtracting the Pluto-only spectrum from the Pluto + Charon spectrum, the infrared data that he and the Riekes had obtained on Mount Hopkins revealed clear evidence for absorption bands characteristic of something never

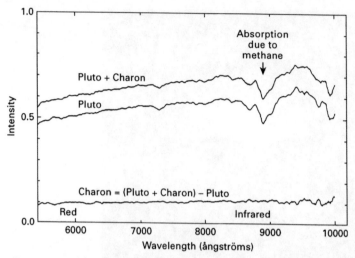

Fig. 3.9: The separate spectra of Pluto and Charon measured by Uwe Fink and Michael DiSanti during the occultation of Charon by Pluto on March 3, 1987. Pluto's spectrum has a strong absorption due to methane near 8900 angstroms, while Charon's does not. Charon's spectrum is flat across the whole wavelength range covered, indicating that Charon's surface has little color. By contrast, Pluto's spectrum gets stronger towards the infrared end of the spectrum, indicating that it is slightly reddish. Note also that the methane absorption feature near 8900 angstroms is not present in Charon's spectrum, revealing that Charon's surface contains little or no methane ice. (Adapted from U. Fink, M. DiSanti 1988, *Astronomical Journal* 95)

before seen in a spectrum of the Pluto system: water ice, and it was on Charon, not Pluto.

Buie, a long-time Plutophile with great skill and intense ambition, was incredibly disappointed at missing the March 4 event, and so he secured telescope time on Mauna Kea for the upcoming April 23 event, the very next one visible at night from Hawaii. Then, along with Dale Cruikshank, and their colleague Larry Lebofsky, Buie tried again. This time, the winds on Mauna Kea were benign, and they obtained a beautiful infrared spectrum that confirmed the presence of water ice on Charon (Figure 3.10).[17]

When all of the groups had published their findings it was clear that the Pluto–Charon pair were no more twins in terms of their reflectivity, color, and composition than they were in their size. Whether their differences stretched back to their origins or instead were due to differences in their evolution, which four billion years of time had

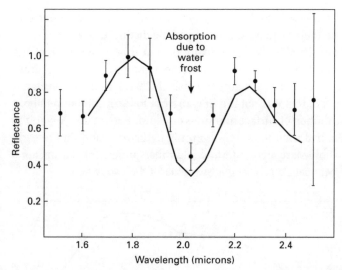

Fig. 3.10: The infrared spectrum of Charon, measured by Marc Buie and collaborators on April 23, 1987. This spectrum was obtained during a total eclipse of Charon, by subtracting Pluto's spectrum from a Pluto + Charon spectrum as described in the text. The measurements for Charon are marked by black dots with vertical error bars through them. The solid line is a laboratory spectrum of water frost. Notice the good correspondence, particularly the dip near a wavelength of 2 microns, which is the characteristic signature of the frost. (The vertical reflectance scale has been set arbitrarily to read 1.0 for the measurement at 1.8 microns.) (Adapted from M. Buie et al. 1987, *Nature* 329)

brought to bear, was not clear, but the main implication was that in just a few hours of observing time spread over March and April 1987, four research groups at four different telescopes had revealed that the dichotomy in size, reflectivity, color, and composition between Pluto and Charon was as striking as the dichotomy between the Earth and Luna.

In the frozen half-twilight of the far outer planetary system where conventional scientific wisdom predicted numbing uniformity, Nature proved it was not listening to conventional wisdom.

The Face of Pluto

Audacious is a good word for what came next. It is one kind of miracle to measure the size of distant, little points of light too small actually to resolve, and it was an even more impressive miracle to uncouple their blended spectra to see the distinct character of each, apart from its close-lying companion. But this! Now some of the eclipse contingent planned to attack the mapping of Pluto, that grain of sand at 1100 meters, using the disk of Charon as their scimitar.

To be fair, this was not the first attempt to learn something about the distribution of surface markings on Pluto. Back in the early 1970s John Fix had tried to do this from his lightcurves and, in the early 1980s, Bob Marcialis undertook a similar project for his master's thesis. Marcialis's project got the better of the two results.

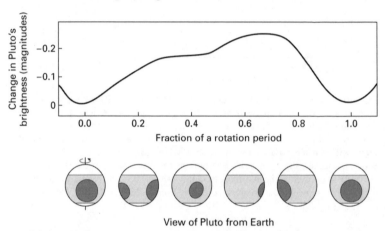

Fig. 3.11: The uneven variation in Pluto's brightness as it rotates on its axis can be explained by the presence of areas on the surface that are darker or lighter than average. (Adapted from R. Marcialis 1988, *Astronomical Journal* 95)

What Marcialis did was to look at Pluto's lightcurve and ask: "What is the simplest distribution of surface markings that would be consistent with Pluto's rotational lightcurve?" This is just the kind of approach astronomers love. It is spare, it is economical, and if Nature holds high the same attributes, then it might even be a good approximation of reality.

Marcialis looked at the lightcurves that Hardie and Walker, and later Andersson and Fix, had obtained, and saw two different dips, one on either side of the lightcurve peak. Marcialis saw that this could be explained by two large, circular dark spots of different sizes on the surface. This was not the only way the lightcurve structure could be explained, but it did fit his data. In keeping with the standard assumption that the change in the lightcurve's amplitude and average value were both due to the increasing amount of equatorial terrain that could be seen on Pluto as it moved around the Sun, Marcialis surmised that the dark spots were near Pluto's equator (Figure 3.11).

To explore this further, Marcialis coded up a computer program to compare the various lightcurves with those that would be produced by Plutos with various bands and spots on them. The trick was to see what combination of spots would produce a lightcurve that best mimicked the actual data. In his computer, Marcialis varied the locations, sizes, and brightnesses of the spots and bands, always searching for solutions that fitted the data with the least number of spots. Marcialis realized that, if the brightnesses, extents, or locations of the surface features had changed with time, then almost any combination of spots was possible. Since there was no compelling evidence for the time variability of Pluto's surface, however, Marcialis assumed that Pluto's surface markings were essentially unchanging through the time span of 30 years over which lightcurve data had been obtained. Marcialis's resort to a Pluto that presented constant surface appearance from the 1950s to the 1980s is an example of a kind of scientific reasoning called Occam's Razor, which is common in research. In effect, the Razor states that, given two or more plausible ways to solve a problem, the approach that makes the least number of assumptions is preferred.

Marcialis's first results were presented at technical talks in 1982 and in his 1983 master's thesis to Vanderbilt University. Marcialis discovered that the decrease in Pluto's average brightness that had been observed over the previous few decades almost certainly meant that the polar regions must be brighter than the equator, which his computer wrapped in a darkened band. This was reminiscent of Walker and Hardie's finding that Pluto's polar regions were brighter than its equator. Superimposed on the equatorial dark band, Marcialis's best-fitting map displayed even darker spots that depressed the lightcurve downward from its peak.

Marcialis's use of rotational lightcurves provided some exciting and tantalizing results, but with the onset of the mutual events it became possible to obtain entirely new kinds of information about the shapes, brightnesses, and locations of Pluto's surface markings. The new technique was to use the passing of Charon's disk in front

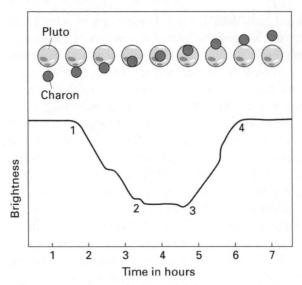

Fig. 3.12: The presence of lighter and darker areas on Pluto's surface caused small bumps in the lightcurve recorded during the mutual events.

of various regions to measure how brightness varied from place to place across Pluto's Charon-facing hemisphere (Figure 3.12).

This kind of modeling is done by putting "tiles" of varying brightness on a computer-generated Pluto, and then blending those idealized surface tiles into a realistic pattern that reproduces the mutual event lightcurves. Like a lot of things we have run across in this chapter, the devil is in the details. For example, once again, there are those pesky shadows of which to keep track. Then there is the fact that Pluto rotates on its axis a few percent during the event, causing a slight change to its brightness. And there is the fact that Pluto is a rounded sphere, rather than the disk it appears to be when projected on the sky. As a result, shadows are curves, and spots and other features can appear to be foreshortened near the limb. And what if Charon has markings too? And is not Charon rotating? Will not this affect the result? Yes, so to be sure of accurate results, the appearance of Charon also has to be modeled. And, just to show the real knottiness of the game, the exact radii of Pluto and Charon must be known, but these are derived from the same data, so, in effect, the model has to solve for the surface markings and radii simultaneously.

Fig. 3.13: Marc Buie (left) and David Tholen in about 1985. Buie and Tholen have been intensively involved in Pluto studies since the early 1980s, and took a major role in the observation and analysis of the mutual events. (Courtesy Marc Buie)

The list of complicating factors that have to be taken into account by such a computer model goes on, but we will not.

No one said it would be easy to map that little rotating grain of sand a kilometer away by using another rotating grain of sand, about half its size, passing in front of it every 6.4 days. But still, if the programming could be done, what a prize these maps would be!

The researchers who took on this work literally put thousands of hours into the details of computer coding, with painstaking reduction of data from the mutual events, and careful analyses of the uncertainties in their results.

Marc Buie and Dave Tholen (Figure 3.13) along with Keith Horne, all fresh postdoctoral workers at the time, undertook one of the most careful mapping projects. PhD student Eliot Young (Figure 3.14) at the Massachusetts Institute of Technology (MIT), and his young PhD advisor, Rick Binzel, undertook another. The two groups each published their findings in 1992 (Figures 3.15 and 3.16).

Buie and colleagues used a combination of mutual event lightcurves and rotational lightcurves stretching back to 1954 as inputs to their

Fig. 3.14: Eliot Young (photo ca. 1995), who as an MIT graduate student worked with his PhD advisor, Richard (Rick) Binzel, to produce a surface map of Pluto based on lightcurve data from the mutual events. (Courtesy Seth Shostak)

model. Young and Binzel relied solely on mutual event lightcurves, which limited their results to the Charon-facing hemisphere of Pluto. Why this limitation? Because Charon's orbital period is known to be identical to Pluto's rotation period, Charon hovers over one place on Pluto. As a result, all of the eclipses of Pluto by Charon were over the hemisphere centered on this location. In contrast, Buie's team used both rotational lightcurves and mutual event lightcurves, which allowed them to map both hemispheres of Pluto.

The two groups applied different mathematical algorithms to determine the distribution of markings on the surface, and used different datasets as inputs to their models. When all was said and done, the two maps that were obtained contained both similarities and contradictions.

Despite their differences, the maps produced by these teams (and the maps produced by several other groups that tried their hand at

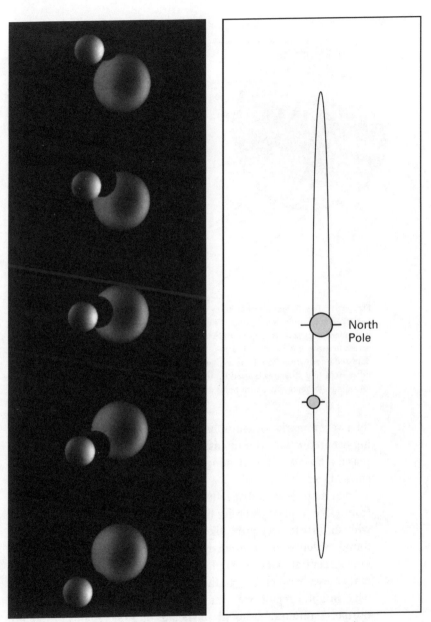

Fig. 3.15: A computer-generated simulation of a transit of Charon across Pluto on February 6, 1988. This simulation, made by Keith Horne, Marc Buie, and David Tholen, resulted from the analysis of lightcurves from previous mutual events. Horne and colleagues used these to construct surface-feature maps that would produce the observed lightcurves. The series of images runs from top to bottom. North on the sky is towards the top and Pluto's north pole is tipped over 90 degrees towards the right, as shown in the schematic on the right. (As shown to a meeting of the American Astronomical Society, January 1989.)

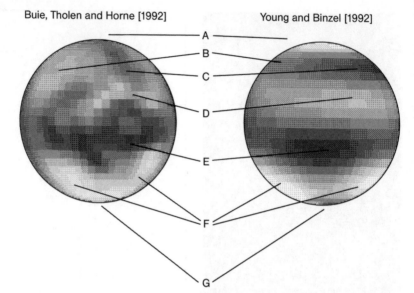

Buie, Tholen and Horne [1992]　　　　　　Young and Binzel [1992]

A

B

C

D

E

F

G

Fig. 3.16: Two different attempts to construct surface maps of Pluto's sub-Charon hemisphere from lightcurve data. The two are based on different techniques and independent data sets. Young and Binzel utilized data taken during transits of Charon across Pluto during the mutual events. Buie, Tholen, and Horne used both mutual event data and information from Pluto's rotational lightcurve. Though there are clearly some differences (e.g., the area labeled B), there are nevertheless some striking similarities, including a large, bright, south polar cap (F). (Courtesy Eliot Young)

this) were remarkable triumphs. Through careful observing, painstaking data reduction, and intensive computer modeling, the maps revealed a number of interesting attributes, many of which HST later showed were in fact real.

Each map showed that reflectivities on the surface varied greatly from place to place, ranging from a low of about 15% (rather like a whitish dirt) to over 70% (like a bright, fresh snow). Each group found evidence for an extensive, bright southern polar cap, with a kind of central, dark notch. Both also found an almost featureless, darker area between the southern cap and the equator, and a somewhat brighter region with interesting longitudinal structure at low northern latitudes. Although the details of these regions differed in the two maps, it is fair to say that south of about 45 degrees north the two groups reached similar findings about the gross distribution of bright and dark regions. Poleward of 45 degrees north, however, there was major disagreement. Young and Binzel's modeling predicted a tiny but bright northern cap. Buie, Tholen, and Horne's computer model predicted a less bright, but more extensive northern cap.

Which map was correct? Which was wrong? Both groups agreed it was not that kind of contest. For one thing, each group used different datasets that emphasized different parts of the Charon-facing hemisphere. Further, because each model involved over 20 variables that had to be simultaneously optimized, each group recognized that neither result represented a perfect solution. Still, despite their uncertainties, both the value and the technical tour de force of the mutual event maps were unquestioned.

The maps revealed a planet that seemed to have at least one and possibly two polar caps, various dark bands and bright spots, and some of the starkest combinations of bright and dark terrain in the solar system.

4
The Importance of Snow

"If we knew what we were doing, it wouldn't be research."
– A. Einstein

"Why is Pluto bright?" It is a simple question, which, in hindsight, has a simple answer. But until 1988, when a scientific paper with this rather unscientific title appeared, no one had ever made the point that Pluto's snow-white surface reflectivity ultimately implies it always, or at least frequently, has an atmosphere.

This is not to say that an atmosphere around Pluto had not previously been suggested. The stream of papers relating to a possible atmosphere on Pluto began in the 1940s, with a landmark work by the legendary planetary astronomer Gerard P. Kuiper. Dutch-expatriate Kuiper was the director of McDonald Observatory, and about the only full-time planetary scientist working in the USA during World War II. Kuiper got the idea to undertake a spectroscopic survey of all of the major satellites of the planets, in order to determine which, if any, had atmospheres. In those distant days before even the rudiments of modern planetary science were known, this project was ambitious in both its scope and its vision.

An atmosphere around a satellite? How intriguing. How bold. How novel. Could satellites even have atmospheres? Well, why not? A planet's atmosphere can be generated in one of several ways: by capture from the gas orbiting the Sun when the planet forms, by outgassing or volcanic activity arising from the interior of the planet, or by vapors given off from surface volatiles. There did not seem to be a reason why a satellite could not generate an atmosphere the same way. But there was still the important detail of whether a satellite could retain its atmosphere against escape to space. If the atmosphere could escape, then it would have to be regenerated again and again and eventually the supply of gas on or in the satellite could run out. This is exactly what would happen to a block of ice in Earth orbit, and what does in fact happen to comets when they are active.

Pluto and Charon, S. Alan Stern and Jacqueline Mitton
Copyright © 2005 WILEY-VCH Verlag GmbH & Co. KGaA, Weinheim
ISBN: 3-527-40556-9

For an atmosphere around a planet or satellite to be stable over eons, the planet's or satellite's gravity has to be strong enough to prevent much escape. Lots of gravity requires lots of mass, so Jupiter, with almost 320 times the mass of the Earth, is very good at holding an atmosphere. But Jupiter is not alone in retaining an atmosphere. The other giant planets are also good at retaining atmospheres, which is in fact part of the reason why they are gas giants. Earth, Venus, and Mars, though far less massive than Jupiter, are able to retain atmospheres as well.

When Kuiper began his search for atmospheres around planetary satellites, all of the planets except Mercury and Pluto had known atmospheres. As for the planetary satellites, however, none seemed to display obvious clouds or other signs of meteorology. And the only satellite we could really inspect closely at the time, Earth's Luna, was clearly devoid of any substantial atmosphere: if Luna ever had a substantial atmosphere it had long since escaped to space.

But what about the satellites of the other planets? Just because clouds and storms were not seen did not mean there were no atmospheres. After all, our instruments might have missed these signs from so far away, or the atmospheres might be clear and not have clouds. So, Kuiper asked himself whether it was physically plausible that some of the satellites of the outer planets might have atmospheres. Even in the early 1940s, several of these moons were known to be more massive than Luna; it was known then, in fact, that Saturn's Titan might even be larger than Mercury. After some simple calculations, Kuiper concluded that it was possible that the large satellites of the giant, outer planets could gravitationally retain atmospheres, particularly since they were cold. Cold helps, because the atoms and molecules in cold atmospheres have less energy than those in warm atmospheres, which makes escape more difficult.

Kuiper was one of those pioneering planetary observers who did not need much motivation to simply go out and look for something new. So if atmospheres might be stable around the cold satellites of the outer solar system, then he would just have a look to see if these atmospheres actually existed.

Given that there was no evidence for outgassing from the interiors of these satellites, or any solid evidence for ices on their surfaces, this was a bold and speculative undertaking, but Kuiper had the resources to carry out a search, and he decided the potential rewards were well worth the effort.

Kuiper correctly surmised that a good way to look for atmospheres in the cold outer solar system was to search for spectroscopic absorptions due to gaseous methane. Two important rationales for methane in particular are that methane is a cosmically abundant molecule and that it remains a gas even at the very low temperatures prevailing in the outer solar system. But best of all, even a little methane creates a huge absorption feature in a planet's or satellite's spectrum, so it can be detected even in trace amounts – something true of few other gases. If the satellites had atmospheres, even thin ones, Kuiper bet, they might show detectable methane absorption signatures.

So, in the winter of 1943–1944 Kuiper used the 82-inch reflector at McDonald Observatory to search for methane absorptions in the spectra of the ten largest planetary satellites, and Pluto (which he included almost as an afterthought). The 82-inch telescope was one of the very largest telescopes of the time, and Kuiper had direct access to it as the director of McDonald. Because photomultiplier tubes (PMTs) and charge-coupled devices (CCDs) were not then available, astronomical spectra were recorded photographically. By today's standards, it was the dark ages, though of course it did not seem that way at the time.

When Kuiper's project was complete, he had made three interesting findings. First, Saturn's largest satellite, Titan, very clearly registered a methane signature. Because Titan's temperature was calculated to be above the 77 K freezing temperature of methane,[18] Kuiper concluded he had discovered the first evidence of an atmosphere around a moon. Second, Kuiper found that Titan was something of an anomaly. None of the other satellites that he had studied showed a clear methane signature. Apparently, atmospheres of satellites were either rare, or at least below his detection limits. Kuiper correctly surmised that the ability of gases to escape easily from satellites was probably an important factor in the lack of atmospheres around most. Third, Kuiper discovered that his equipment was not sensitive enough to explore the methane question with regard to two of the faintest targets at which he looked: Neptune's moon Triton, and Pluto. Being conservative, Kuiper did not say Triton and Pluto did not have detectable methane absorptions. His data just were not good enough to make definitive conclusions. So what Kuiper did was to place neither a "yes" nor a "no" in the atmosphere column for these two distant bodies. For each, he just entered "?" in his roster.

The Low Roads

Decades passed between Kuiper's 1944 paper and the dramatic and unequivocal discovery of methane ice on Pluto in 1976 by Dale Cruikshank (who, we note, had been one of Kuiper's later students). But that raised the question of whether Cruikshank and colleagues' discovery of methane ice really meant Pluto has an atmosphere.

The situation at Pluto was not as clear as the situation at Titan, where the surface temperature was above the freezing point of methane, and methane could thus easily sublimate from the ice to form an atmosphere. But because Pluto is so far from the Sun, and therefore so weakly warmed by it, Pluto's surface temperature is well below the freezing point of methane. As a result, the methane Cruikshank detected would largely remain frozen. The operative question, then, was how much, if any, of the methane would sublimate to gas, and how much of Cruikshank's absorption signature was due to methane gas.

This was not an easy question to answer. For many substances, a scientist with such a question could just enter into his computer the published laboratory spectra of the gas and solid and use these to find out how much gas and how much solid ice he or she needed to match the observed spectra. (This technique in its simplest form is akin to mixing paints of primary colors to match a given hue.) The problem with methane, however, was that the library of *low-temperature* methane spectra was not complete. After all, why would anyone have thought to measure the spectrum of methane ice and gas just above absolute zero?

Worse, from the little data that did exist back in the late 1970s, there did not seem to be clear distinctions between a solid and a gas unless observations were made at very high spectral resolution. So, the goal became to get an infrared spectrum of Pluto at high spectral resolution. This is easily said, but the telescope and detector technology to do this was not available in the 1970s or the 1980s. In fact, it was to be 1994 before this was accomplished.

So what did the ardent astronomical explorers of the wild black yonder of the 1970s and 1980s do in the meantime? They did what astronomers commonly do. Given a rock-solid roadblock like this, they resorted to less direct but more clever means to tackle the obstacle.

Theorists attacked the problem by simply calculating how much gas methane ice would generate at Pluto's temperature. To make this calculation, one simply had to look up for methane ice something called the vapor pressure formula, put it into a computer, and plug in the temperature. The result: a predicted atmospheric pressure! Unfortunately, vapor pressure relations are exponentially (i.e., very, very, *very*) sensitive to the assumed temperature of the ice. In fact, in the temperature range expected for Pluto, the methane vapor pressure *doubles* every time the temperature of the ice increases by just two degrees. So, to get a meaningful result, it was first necessary to have an accurate surface temperature for the methane ices on Pluto.

And what is the surface temperature of Pluto? Well, because the technology was not available in the late 1970s and early 1980s to measure the feeble heat emitted from Pluto across the five billion kilometers that separated Pluto from Earth, the temperature also had to be calculated. That depended on knowing how well the surface reflects light and emits heat, neither of which was known at the time. So the theorists made their best guesses, started calculating, and concluded that the surface was most likely between about 40 K (–390 °F) and 65 K (–340 °F). But whether the answer was 40 K or 65 K made an enormous difference in the expected pressure – a factor of about 1000!

Things got even worse when University of Texas astronomers Larry Trafton and Ed Barker pointed out that the methane might not be pure, and that the amount of atmosphere that Pluto's frozen methane would generate would depend on its purity. Clearly, the vapor pressure calculations were not good enough – there were just too many uncertain variables that went into calculating a result.

A second attack was mounted by observers (who generally distrust theoretical results anyway). Recall that back in 1970, John Fix and his collaborators had obtained that first, tentative, 21-point spectrum of Pluto revealing its red color, and a possible little upturn on the blue (i.e., left) side of the spectrum (see Chapter 2). Fix and co-workers suggested at the time that the upturn might be due to Rayleigh scattering, the effect that makes the Earth's sky blue.

The operative thing about Rayleigh scattering is that it takes a whole lot of atmosphere to generate the bluing seen above our heads every day (if it did not, you would see a blue roomful of air in every auditorium). What this means is that, if Rayleigh scattering was indeed the cause of the upturn in Fix and colleagues' spectrum, then Pluto had to have quite an atmosphere. Of course, you already know what Fix and colleagues did not: that the little blue upturn in their spectrum was an instrumental artifact. But no one knew that in the late 1970s, so as a result a whole series of theoretical speculations were launched about this upturn.

Fortunately, before too many theorists wrote papers about the implications of Fix's blue spectral upturn, McDonald Observatory astronomers Ed Barker and Bill and Anita Cochran obtained a far better spectrum of Pluto in 1979 showing that Fix and colleagues' result was spurious. The implication that there was no strong Rayleigh scattering in Pluto's atmosphere was clear: if there was an atmosphere around Pluto, it had to be pretty tenuous, certainly less than Mars's atmosphere – which itself is less than 1% of that of Earth's.

What other means might there be to get at the question of whether Pluto has an atmosphere? Well, if there had been an opportunity to fly one of the Voyager spacecraft to Pluto, all of this could have been easily wrapped up using its onboard instruments to study Pluto at close range.[19]

Similarly, if Pluto's atmosphere had generated variable hazes or clouds, the atmosphere could have been detected by the ever-changing, day-to-day variations that their presence would have induced on the lightcurve. But Pluto's lightcurve showed no such variations, from which some concluded there was either no atmosphere, or if there were an atmosphere it was very clear and lacked the kind of variable meteorology Earth displays.

All these twists and turns are fun to enjoy from the easy-chair that hindsight provides, but in the 1970s and 1980s it was not fun: it was difficult and frustrating. People were doing their best with the limited tools of their time to glimpse the true nature of Pluto. But the tools were just not up to the task at hand, given that Pluto's atmosphere, if real, was known to be so rarified.

A Clean Machine: The Road Less Traveled

All the frustration was to come to an end in the summer of 1988, when the issue of Pluto's atmosphere would be settled definitively. However, before we see what was discovered in 1988, there is one more tale that is worth recounting.

It began with the same question that we posed at the beginning of this chapter: *Why is Pluto Bright?* By all rights, Pluto should not be bright, because the very methane that lies on its surface should, over time, react with ultraviolet sunlight to create more complex carbon- and hydrogen-bearing (i.e., hydrocarbon) molecules, such as ethane, ethylene, and acetylene. These more complex hydrocarbons are dark, whereas methane is bright. As a result, methane turns into a blackish-reddish residue if left untouched and unshielded from ultraviolet sunlight.

The fact that methane would react and turn darker had been firmly established in laboratory experiments replicating conditions in the outer solar system. These studies indicated that the time for exposed methane on Pluto's surface to turn black was just a hundred thousand years – a veritable wink of the eye compared to the age of the solar system.

So how could Pluto's methane stay bright? There were really only three viable alternatives, laid out in a research paper called "Why is Pluto Bright? Implications of the Albedo and Lightcurve Behavior of Pluto":[20]

- Pluto could have an atmosphere that shields the methane from the ultraviolet rays that would darken it;
- Pluto could have volcanoes that routinely repaint its surface with fresh, bright methane snow;
- Pluto could have an atmosphere that cleans and repaints its surface with fresh, bright methane snow.

All three of the alternatives implied an atmosphere of one sort or another, but the atmospherically driven surface repainting made the most sense. In part this was because it was the simplest explanation, and in part it was because it did not invoke any new claims about Pluto the way volcanoes did. Instead it was just a byproduct of Pluto's orbital cycle that offered a natural explanation for a long-standing mystery about Pluto's changing lightcurve.

The gist of the atmospheric cleansing scenario is this. Regardless of the present temperature on Pluto, whether it is 40 K or 65 K, there would be some methane vapor, and that would make an atmosphere, even if it were only very tenuous. After each Pluto perihelion, the temperature would start to drop as Pluto draws farther and farther from the Sun and cools. And as the temperature drops, the exponential dependence of the methane vapor pressure on temperature would cause the amount of methane in the atmosphere to decrease rapidly. Calculated estimates showed that the atmospheric bulk could drop by factors of 100 to 1000 within 20 years after perihelion.

But where would the atmosphere go? It would condense out on the cold surface, as a frost layer perhaps a few millimeters to a few centimeters (0.1 to perhaps 2 inches) deep. Since methane frost is bright, Pluto would brighten after each perihelion as the planet receded from the Sun and cooled. Then, for about two centuries, the atmosphere would lie there, frozen as a snow on the surface while Pluto glided out to its farthest point from the Sun, rounded its orbital bend, and again approached the relative warmth of its next perihelion. All the while, of course, the fresh methane snow would be subject to ultraviolet sunlight, which would begin to darken it. But then (this is the best part), as the surface warmed again when Pluto approached its next perihelion, the methane would sublimate (i.e., it would be converted from ice to gas in the same way that a block of dry ice "smokes"). This sublimation would regenerate the atmosphere, but it would leave behind the inert byproducts of ultraviolet processing. In this way, the methane snows created two centuries before would disappear from the surface and, as they did so, they would leave behind their darkened, less volatile residue, something like the way that a snowfall darkened by road salts evaporates by solar power and leaves behind its dirty residue (Figure 4.1).

Each orbit, the same cycle should occur. As perihelion approached and the atmosphere formed, Pluto would darken due to the removal of the overlying burden of snows that had condensed onto the surface two centuries before. If the snows were not entirely uniform over the planet, then as they were converted back into the atmosphere by heating, the thinnest snows would probably disappear first, making Pluto look spotty as it approached perihelion. After perihelion, as the atmosphere once again cooled and condensed back onto the ground, the surface would brighten with the freshly fallen snow. In effect,

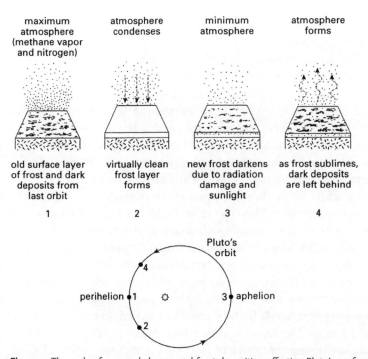

maximum atmosphere (methane vapor and nitrogen)	atmosphere condenses	minimum atmosphere	atmosphere forms
old surface layer of frost and dark deposits from last orbit	virtually clean frost layer forms	new frost darkens due to radiation damage and sunlight	as frost sublimes, dark deposits are left behind
1	2	3	4

Fig. 4.1: The cycle of seasonal change and frost deposition affecting Pluto's surface. The top part shows the effect on the surface of the seasonal changes. The lower part shows the approximate position of Pluto in its orbit corresponding to each stage, though it is not possible at present to predict the exact timing. (1) As Pluto reaches perihelion, increased heating by the Sun causes nitrogen and methane to sublimate from the surface into the atmosphere. The surface becomes darker as a photochemical sludge, which does not sublimate, is left behind. (2) As Pluto's distance from the Sun increases after perihelion, and its surface temperature falls, nitrogen and methane condense out of the atmosphere onto the surface, creating a fresh, bright layer of new frost. (3) The methane frost on the surface darkens in color as ultraviolet radiation from the Sun causes chemical changes to take place. (4) The atmosphere starts to re-form as Pluto's orbit carries it closer to the Sun and the surface grows warmer again. The seasonal cycle repeats.

the atmosphere would act as a laundry, cleansing the uppermost layers of the surface every 248-year orbit and, in doing so, keep Pluto bright. Since only a few centimeters of UV-processing residue would be created over the age of the solar system, there did not seem to be any problem with the methane below the residue wafting through it when heated at perihelion. In short, if the supply of methane near

Pluto's surface was fairly plentiful, the atmospheric laundry could go right on for eons.

The thing that made this work so intuitively believable was not that the numbers behind the story worked perfectly, but that the physics was so simple and logical. Apparently, without realizing it, observers had already been seeing the sublimation of the last orbit's snows into the atmosphere, exposing the residue layer below. For it was the orbital cycle of atmospheric collapse, dormancy, and rebirth that offered a wonderfully consistent explanation for the long-perplexing mystery of why Pluto's reflectivity had been declining ever since Bob Hardie and Merle Walker's 1950s measurements! There was even some evidence that Pluto had grown redder as it grew darker, which matched with the atmospheric laundering idea as well.

So in 1988, when "Why is Pluto Bright?" appeared, the response among Pluto researchers was very positive. The whole question had been overlooked during the dozen years since methane had been discovered. But once the issue was exposed, there just did not seem to be an easier explanation for how a methane-coated Pluto resisted blackening in the Sun than the atmospheric laundry cycle. Atmospheric laundering seemed to explain a lot that had not fitted together before, and so it captured the stage and convinced most people that we should not only expect an atmosphere on Pluto, but a dynamic one at that. The operative questions thus moved from, "Is there an atmosphere?" to, "How much of an atmosphere would Pluto generate, and how it could be definitively detected?"

The High Road, At Last

Although the logical framework convinced many planetary astronomers that Pluto's methane signature certainly did imply that Pluto had an atmosphere, observers insisted that nothing substitutes for direct evidence. And that is right. So detect it they did.

The clear and unambiguous detection of Pluto's atmosphere occurred on a June day in 1988. And as you might expect, it did not come without surprises.

The definitive detection of Pluto's atmosphere was made by observing a stellar occultation by Pluto. A stellar occultation is an astronomical event in which a planet (or asteroid or planetary satellite) temporarily moves in front of a star. If a planet has an atmosphere, even one a million times less dense than Earth's, then that atmosphere will refract starlight passing through it, thereby imprinting an unmistakable signature that reveals the presence of the planet's atmosphere.

Thus, a very good test for whether or not Pluto (or any other body) has an atmosphere is to observe it occulting a star. A star occulted by an atmosphere-less world (like the Moon) would disappear essentially instantaneously as the limb of the planet quickly cut off the light of the star. However, if the world occulting the star sports an atmosphere, refraction causes the signal to decline more gradually. How gradually depends on the density and the thickness (which astronomers like to characterize by something called the "scale height") of the atmosphere. So in an occultation observation, the experimental objective is to obtain a running record of the brightness of the star as it passes behind the planet and to determine whether it just winks out in a tiny fraction of a second or instead slowly fades, perhaps over tens of seconds, or even minutes.

Stellar occultations have been used to probe planetary atmospheres for decades. As we pointed out, however, Pluto's small size and great distance conspire to make it a tiny dot on the sky, and a slow moving one at that, so stellar occultations by Pluto are very rare.

The first real attempt to observe an occultation by Pluto occurred when Ian Halliday, Bob Hardie, and Otto Franz set out in 1965 to determine Pluto's diameter, which was at that time almost a complete unknown. Although this talented trio of astronomers did not intend to search for an atmosphere, they could have detected one with their equipment. Unfortunately, the track of Pluto's shadow on the Earth due to the occultation could not be predicted very accurately in those days, so Halliday and team were in the wrong place and all they observed was a tantalizing near-miss. As seen in their telescope, the hoped-to-be-occulted star passed a maddening ten-thousandth of a degree (corresponding to 3300 kilometers, or 2060 miles) off Pluto's limb: close, but no diameter and no information on an atmosphere.

The next observed Pluto occultation came in April of 1980. By this time it was clear that an occultation presented an opportunity to search for Pluto's possible atmosphere but the event predictions indicated that most of the shadow track would lie over the south Atlantic, in daytime, with only a slight chance that observers at the south end of Africa might be able to catch the event at night. This was of course the event Alistair Walker luckily observed (as we described in Chapter 3), but from Walker's vantage point the shadow was cast by Charon instead of Pluto. Although he did not achieve his intended goal of observing an occultation of Pluto, he was lucky enough to catch Charon's shadow in his telescope and accurately estimate its diameter for the first time.

A third promising event occurred in 1985 but again the shadow track fell almost entirely over open water. In fact, the only observers anywhere near the shadow track who were aware of how important such an observation was to studies of Pluto were Noah Brosch and Hiam Mendelson at the Wise Observatory in northern Israel. As luck would have it, a great Israeli–Jordanian air war was taking place overhead that night, replete with artillery fire, air-to-air rockets, cannon shots, and tracer rounds. Almost unbelievably, Brosch and Mendelson managed to observe the event despite the circumstances overhead. But when they inspected the data they obtained, they saw that the faint signal from Pluto and the target star was marred by bright flashes from the air battle that had been taking place overhead. It was (and still is) probably the only astronomical observation ever made through a sky filled with dog-fighting Mig-23s, F-4 Phantoms, and F-15 Hornets. Although Brosch and Mendelson believed they could see telltale evidence for refraction in the way the star disappeared, the data were so contaminated with light from air battle artillery and tracers that others who inspected the data just could not say whether there was something definitive there or not. Perhaps the two Israelis had detected something indicating an atmosphere, or perhaps not. Yet another try would be required.

That attempt came in 1988, when predictions indicated that yet another star would be occulted by Pluto, this time at night over a wide swath of land and sea. The star, which went only by the simple provisional name "P8" for Pluto occultation candidate number eight, was wholly inconspicuous except for its appointed date with Pluto. And when it became clear that P8 looked like it would actually pass behind Pluto (rather than off its limb), two teams of astronomers went out to capture systematically as much data as possible, and resolve the issue of Pluto's atmosphere once and for all.

For months before the June 9, 1988 event, planning was under way. Bob Millis (Figure 4.2) and Larry Wasserman of Lowell Observatory,

Fig. 4.2: Robert Millis, the present-day director of Lowell Observatory. Millis and his colleague Larry Wasserman led the planning for ground-based observations when Pluto occulted P8 in June 1988. Photo ca. 1995. (Courtesy R. Millis)

and Jim Elliot (Figure 4.3) and Ted Dunham of the Massachusetts Institute of Technology (MIT) knew the stakes were high and together they organized an observing campaign that included at least seven telescopes.

The event was predicted to cast Pluto's shadow over the south Pacific, Australia, and New Zealand. Knowing that factors like Charon's tug on Pluto made the precise track of the shadow impossible to predict, Millis and Wasserman organized a fence-line of telescopes to snare their prey.

MIT's Elliot, along with his colleague Ted Dunham, organized a wholly separate effort from the one Millis and Wasserman spearheaded. Elliot's team petitioned for and succeeded in bringing NASA's

Fig. 4.3: Jim Elliot, an experienced observer of occultations, successfully led the Kuiper Airborne Observatory expedition to fly over the best predicted ground track for Pluto's stellar occultation in June 1988. Photo ca. 1995. (Courtesy J. Elliot)

flying Kuiper Airborne Observatory (KAO) down to Hawaii in order to fly it out over the predicted ground track, 3000 miles to the south (Figure 4.4). (This unusual move of the KAO to Hawaii from its home base in California was costly, and testified to the scientific value of the highly anticipated 1988 occultation.) The KAO had for years been equipped with a 91-centimeter (36-inch) telescope.[21] For the P8/Pluto occultation Elliot and colleagues placed a high-speed (i.e., rapidly framing) CCD camera called SNAPSHOT behind the telescope; SNAPSHOT had been specially designed for occultation studies.

The KAO offered Elliot two key advantages. First, it was mobile, so it allowed him to change his observing plan and location as the predicted path of Pluto's shadow improved in the days and hours leading up to the event. No ground-based telescope could do that. Second, by flying high in the stratosphere at altitudes of 40 000 feet or more, the KAO could essentially make certain that no matter what happened below, weather would not shut out the observation effort.

Elliot and Dunham had used the airborne KAO to capture many occultations by planets and satellites over the years. Mars, Jupiter, and Io were just a few of their trophies. Their best-known achievement

Fig. 4.4: The Kuiper Airborne Observatory (KAO), which operated from Ames Research Center from the mid-1970s until it was retired from service in 1995. A converted military cargo plane, the KAO carried a 36-inch reflecting telescope for astronomical research. The square telescope aperture in the fuselage can be seen just in front of the wings. (NASA)

had come in 1977, when they discovered that Uranus has rings. Later, when Elliot was asked why he spent so much effort preparing for the 1988 Pluto event, he quietly said, "We wanted to do something hard, and the only thing left was Pluto."

At the same time that Elliot and Dunham's team were preparing to use the KAO, Millis and his team were finalizing their careful plans for their ground-based effort. The logistics took months to plan. Observatories had to be enticed to participate. Mobile telescopes had to be made ready to ship to New Zealand and Australia to fill gaps in the occultation fence line. And new observations of Pluto and P8 had to be made to refine the predicted time and exact track where the event would be visible.

Between Millis and Wasserman's careful, D-Day-like ground-based planning, and Elliot and Dunham's successful bid to fly the KAO

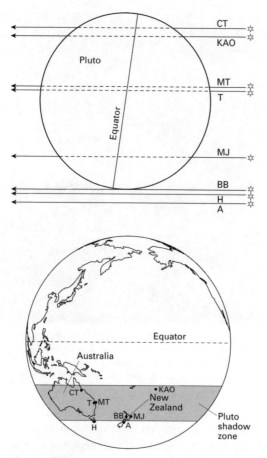

Fig. 4.5: Pluto's occultation of the 12th magnitude star P8 on June 9, 1988 was visible from a range of latitudes covering parts of Australia, New Zealand, and the south Pacific. Observations of the event were obtained from Charters Towers (CT), Toowoomba (T), Mount Tambourine (MT) and Hobart (H) in Australia, from Mount John (MJ), Black Birch (BB), and Auckland (A) observatories in New Zealand, and from NASA's Kuiper Airborne Observatory (KAO) flying over the south Pacific Ocean. The apparent track of the star behind Pluto as seen from each of the locations is shown in the top part; Pluto's shadow track on the Earth is shown in the bottom one.

airplane "down south" to study the event, Pluto was not going to get away this time. And indeed, it did not.

When June 9 came, the equipment worked, the weather cooperated, and the ground track of Pluto's shadow was close to prediction. Seven different telescopes at remote Australian and Tasmanian locations with faraway names like Mount John, Toowoomba, Hobart, and Charters Towers captured the lazy disappearance of P8 behind Pluto (Figure 4.5).

Among the various ground-based telescopes that observed the event, P8's disappearance was seen to take between 70 and just over 150 seconds. In fact, the refraction effect was so strong that some of the telescopes never even saw the signal fully bottom out, as would occur if the star totally disappeared and only Pluto remained.

The KAO saw it too, rewarding Jim Elliot and his stratospheric explorers (Figure 4.6) with 100 seconds of precious occultation data. Aboard the KAO P8's gradual fade to black behind Pluto was met with shouts of joy.

Fig. 4.6: Researchers from the Massachusetts Institute of Technology (MIT) the morning after they successfully used the Kuiper Airborne Observatory to observe the occultation of a star by Pluto on June 9, 1988. Left to right: Amanda Bosh, Stephen Slivan, Leslie Young, Edward Dunham, and James Elliot. (Ben Horita, courtesy Stephen Slivan and Leslie Young)

"What this means," Jim Elliot declared on seeing the data, "is that Pluto has an atmosphere!" By the next day, the news was out in papers around the world. *Sky & Telescope* reporter Kelly Beatty, who had ridden with Elliot and colleagues aboard the KAO, wired in a story to the morning's *Boston Globe*. In a page one headline, Beatty and the *Globe* shouted: "Key Pluto Find!"

Mining the Occultation Data

The stories that ran in the newspapers in the USA in early June of 1988 painted the discovery of Pluto's atmosphere as the end of a quest. But to those who observed the event, this was not an ending at all: it was a beginning.

For it was as if God had shined a searchlight through Pluto's atmosphere, and as the beam from P8 passed closer and closer to Pluto's limb, it delved deeper and deeper into the envelope of alien air surrounding this distant planet. With the data gathered during the stellar occultation it would actually be possible to unravel the *structure* of Pluto's atmosphere. The occultation teams knew that, and they did not waste any time.

By comparing the numerous 1988 occultation datasets that had been gathered, it quickly became clear that some of the lightcurves had reached down to the bottom of the atmosphere (or at least to a layer in it where the light level bottomed out). Other lightcurves had not got that far because the telescopes that obtained them were located off the center track of the occultation, and so had cut through the atmosphere at a more grazing angle near one of Pluto's poles. From the lightcurves that did reach the bottom level, it was determined that the pressure there was between 3 and perhaps 10 microbars (i.e., between 3 and 10 *millionths* of the air pressure at sea level on Earth) (Figure 4.7).

This meant that the gas at the base of Pluto's atmosphere was rarified indeed. In fact, the gas at Pluto's surface is as tenuous as the air 80 kilometers over our Earth-bound heads, on the very edge of space. It was the most rarified atmosphere ever detected by occultation. No wonder it had been so hard to detect with other, less decisive techniques! So tenuous was Pluto's atmosphere that a simple calculation showed that from the surface of Pluto, the sky overhead at noon would look as inky black as space itself.

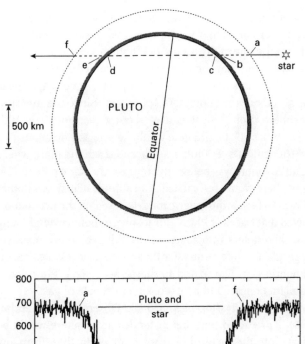

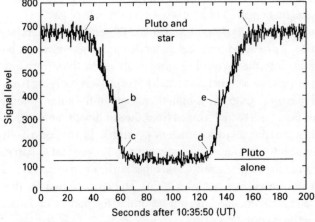

Fig. 4.7: The lightcurve obtained by the Kuiper Airborne Observatory during the occultation of a star by Pluto on June 9, 1988. The upper diagram shows the apparent track of the star behind Pluto. The starlight from P8 was not cut off abruptly, but dimmed gradually, as shown in the lightcurve below. A corresponding effect was seen as the star emerged from behind Pluto. This is a clear signature of refraction, indicating that Pluto has a tenuous atmosphere, which is represented by the outermost circle. The letters a, b, c, d, e, and f show how the progress of the lightcurve is linked to the position of the star relative to Pluto. The declining and rising parts of the lightcurve each display two slopes. When nearest the planet, the star appeared to dim and brighten more rapidly than when it was somewhat farther away. To explain this, there must be some kind of layering in Pluto's atmosphere, which could be a sharp temperature gradient, the presence of a haze layer, or both. (Adapted from J. Elliot et al. 1989, *Icarus* 77)

But that was not all. Pluto's atmosphere was also revealing itself to be different from any other yet explored and it was incredibly interesting. From the rate of the occultation lightcurve's decline, it was possible to determine that the atmosphere extended upward many hundreds of kilometers and that the atmospheric scale height near the surface was about 60 kilometers (38 miles). Since the pressure in an atmosphere drops by a factor of about 3 every time the altitude increases by a scale height, one would have to ascend about 300 kilometers (190 miles) over Pluto's surface to reach the point where 99% of the atmosphere was below. By contrast, the same point in the Earth's atmosphere would be reached at an altitude of just 40 kilometers (25 miles). In fact, atmospheric models based on the occultation results showed that traces of Pluto's atmosphere should extend many thousands of kilometers into space. Thus, whereas more massive, higher-gravity planets like Venus, Earth, and Mars have only a razor-thin atmospheric skin (compared to the planet's size), Pluto looks more like a planet encased in an deep, spherical halo of gas.

Theorists like Larry Trafton of McDonald Observatory and Ralph McNutt at MIT predicted that this distended atmosphere would be escaping away into the vacuum of space more easily than from any other planet. Perhaps as much as a half a ton of gas was escaping from Pluto every second. Over the 4-billion-year age of the solar system, this rate of loss of gas (which comes from the sublimation of surface ices) implies that up to 30 kilometers (as much as 100 000 feet) of terrain on Pluto's surface has slowly but surely been lost to space. Ancient mountain ranges, craters, and other surface constructs could be completely erased to space over time. The term coined for this process was "escape erosion." As was becoming the norm with Pluto, it was something completely new.

Smoke? Or Mirrors?

It did not take long before something else was discovered in the 1988 occultation lightcurves. Closer inspection of the data revealed that the slope, or rate of decline of many of the lightcurves, steepened at about the point where half of P8's light had been extinguished. What would cause the atmosphere to suddenly change the way it affected the probing starlight? Two theories were put forward to explain this curious twist.

Fig. 4.8: Leslie Young, ca. 1996. As a graduate student at MIT in the early 1990s she worked on the properties of Pluto's atmosphere. (J. Mitton)

The first was Jim Elliot's idea. Perhaps Pluto's atmosphere contained a haze layer that extinguished some of the light. Haze layers are fairly common in planetary atmospheres. The Earth's atmosphere, for example, has many thin layers composed of aerosol droplets, ice particles, and dust high above it.[22] Working with his then student Leslie Young (sister of Pluto-mapper Eliot Young) (Figure 4.8), Elliot suggested that the haze could be a natural byproduct of ultraviolet sunlight acting on methane in the atmosphere, just as the same ultraviolet sunlight would darken the surface methane. Elliot and Young calculated that the top of the haze layer would lie 30 or 40 kilometers over Pluto's icy terrain, and would soak up about 35% of the light falling through it to the surface (Figure 4.9).

An alternative to the haze idea was hit upon separately by Von Eshleman at Stanford University and a University of Arizona team consisting of Bill Hubbard, Roger Yelle, and Jonathan Lunine (Figure 4.10). Each recognized that the lightcurve steepening did not *require*

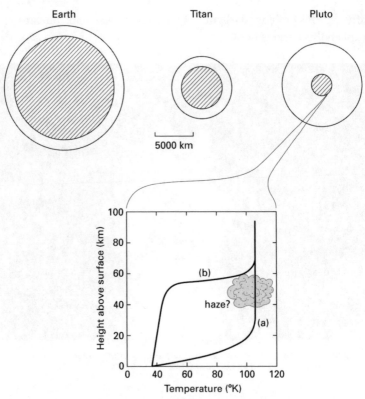

Fig. 4.9: Top: The extents of the atmospheres surrounding Earth, Titan, and Pluto are shown to scale for comparison. The limits shown represent in each case the upper stability limit, known as the *exobase*. Pluto's tenuous outer atmosphere is huge compared with the small planet. (Adapted from a diagram by Roger Yelle.) Bottom: Observations made during the 1988 occultation of the star P8 by Pluto probed a region of Pluto's middle atmosphere a few tens of kilometers thick (solid lines). They suggest either a layer of haze (corresponding to line a), or a steep increase in temperature with height (corresponding to line b). A combination of the two effects might also be at work. The lines are extrapolated to the surface. In reality, there is some uncertainty (of the order of 25 kilometers) about the zero point of the altitude scale. An average value is used for the purpose of this schematic illustration. (Adapted from a diagram by Leslie Young)

any haze at all. It could instead be created by an increase in temperature with height in Pluto's near-surface atmosphere. Further, the Hubbard, Yelle, and Lunine team found that this heating could naturally result from the presence of atmospheric methane (which makes a kind of greenhouse effect). If there was enough methane in the atmosphere, and nothing to cool it, then the increase in temper-

Fig. 4.10: Roger Yelle (left) and Jonathan Lunine in 1996. Like Von Eshelman, they argued that the stellar occultation data could mean that the temperature of Pluto's atmosphere increases with height above the surface. (Photo A. Stern)

ature with altitude would change the amount of refraction in just the right way to create the lightcurves that had been observed by the KAO and the various ground-based groups. Yelle and Lunine calculated that, if their model were correct, the upper atmospheric temperature high above Pluto's surface would reach 106 K (–270 °F) or so, far warmer than the surface temperature (though still frigid by anyone's standards).

So, was it a haze or a thermal effect that steepened the occultation lightcurves? We still do not know. The 1988 data can be fit by either model. In fact, the two explanations are not mutually exclusive: there could be both some haze and some temperature rise with altitude.

Interestingly, however, both the haze and "hot" atmosphere interpretations call into question one of the results obtained from the mutual events: the radius determination for Pluto. The best mutual event estimates give radius values between about 1145 and 1170 kilometers (715 and 730 miles). But if there is a haze, then Charon could have disappeared just before it sank behind Pluto's limb and the true surface could be somewhere in the murky depths, 10 or 50 or even 75 kilometers below where it seemed to be. Just as plausibly, the thermal gradient model predicts that a kind of optical mirage could be created by the atmosphere, which could again have caused Charon to disappear from view before it actually sank below Pluto's limb. If the occultation lightcurves did not actually get down to the hard surface, but simply to a haze deck or mirage-like level, then the atmosphere would extend further down and the surface pressure would be higher than a few microbars. Indeed, the pressure on the hard surface could be as great as about 50 microbars.

This issue of Pluto's true radius is of more than academic interest, at least to academics, because packing the same mass into a smaller volume implies a denser planet. A denser planet, in turn, would mean that Pluto contains even more rock, and less ice, than previously thought.

Taking Pluto's Temperature

One of the most interesting things about the atmospheric models developed from the stellar occultation data was that they suggested some unexpected problems with our understanding of something else very fundamental about Pluto: its surface temperature.

Any first-year physics student with a hand calculator could figure out the surface temperature *expected* on a Pluto being warmed by sunlight. The formula for this is called the radiative equilibrium temperature relation, and it is not complicated. All you have to know is the surface reflectivity (that is, how much sunlight is reflected away and is therefore not available to heat the surface), the efficiency at which the surface emits heat (often assumed to be essentially perfect, i.e., near 100%), Pluto's distance from the Sun, the Sun's luminosity, and Pluto's atmospheric pressure. The mutual events had generously contributed the reflectivity. Pluto's distance from the Sun and the Sun's luminosity can be calculated, or can be looked up in a book. The surface pressure, which the occultation results provided, is needed because it tells you how fast the ice on the surface is sublimating, which robs some of the energy that would otherwise go into heating the surface. An everyday consequence of this sublimation cooling

Fig. 4.11: Mark Sykes, one of the young Plutophiles, in about 1996. Using data from the Infrared Astronomical Satellite (IRAS), Sykes made one of the first reliable estimates of Pluto's surface temperature. (Courtesy Mark Sykes)

effect is the chill that comes from alcohol-based products (like after-shave lotion) evaporating on your skin. Loading all these factors into the calculator, our student would predict a temperature between about 50 and 60 K (–370 and –350 °F) on Pluto's surface.

Comfortingly, when the European Space Agency and NASA launched an infrared observatory called IRAS into space in 1983, it obtained thermal measurements of Pluto that were interpreted by a talented planetary scientist named Mark Sykes of the University of Arizona (Figure 4.11) to give a temperature of 55 to 60 K.

So what is the problem? The rub came from a mismatch between the pressure measured from the stellar occultation and the pressure resulting from sublimating methane at 55 or 60 K. The occultation data implied that even if the atmosphere was about as deep as it could possibly be, the surface pressure could not exceed about 50 microbars. But if the surface temperature of Pluto's ices were 55 or 60 K, then the predicted pressure resulting from sublimating methane would be a hundred times higher than that! There was no way that the pressure measured from the stellar occultation lightcurve and surface temperature information could be made to correspond.

Fig. 4.12: The 30-meter diameter radio telescope of the Institute de Radio Astronomie Millimetrique (IRAM), located atop the Sierra Nevada mountains near Granada, Spain. IRAM is operated jointly by France, Germany, and Spain. The telescope weighs over 800 tonnes. The building beside the telescope houses the control center, offices, and a dormitory where astronomers stay when working at the telescope. Using this telescope in 1988, a team of European astronomers headed by Wilhelm Altenhoff obtained the first evidence that much of Pluto's surface is colder than IRAS had indicated. (Ron Allen, STScI)

Something had to give and the first hint of what that might be came in 1986 from an unexpected quarter. For it was then that a team of Europeans headed by Wilhelm Altenhoff of the Max Planck Institute for Radio Astronomy in Bonn obtained a different kind of temperature measurement of Pluto. This measurement was obtained using the spanking new, high-tech IRAM (Institute de Radio Astronomie Millimetrique) radio telescope perched atop the Sierra Nevada range in the south of Spain (Figure 4.12). Altenhoff and colleagues found something very different from what the IRAS data had apparently shown. *Their* data indicated that Pluto's surface was probably a lot colder than 60 K – more like more like 40 K (–390 °F). But it's a dry cold.

Hardly anyone in the US planetary research community knew about or appreciated the importance of Altenhoff's result. Why? For one thing, as a good Euro-citizen, Altenhoff published his findings in a European astrophysics journal which few Plutophiles (all of them American) read. Further, of those who did know about this odd result, essentially all discounted it because it was at odds with the then common viewpoint that the temperature on Pluto "must be" near 60 K, the temperature that a methane surface would obtain at Pluto's distance from the Sun. Further, at the time Altenhoff published his work, the 1988 occultation of P8 had not occurred, so there was no atmospheric pressure measurement to give pause to the warmer, IRAS-derived temperature. So Altenhoff's prescient hint of something awry with our conception of Pluto lay fallow in the technical literature.

Meanwhile, theorists like Yelle and Lunine were pointing out something equally important that eventually came to bear on this puzzle. If their warm upper atmosphere interpretation of the occultation results was correct, then it implied something other than methane had to inhabit Pluto's atmosphere in order to fit the measured 60-kilometer (38-mile) scale height. Since methane gas would produce a much larger scale height than the data apparently allowed, the suspected second constituent had to be heavier than methane, and at least as abundant. What could that "second gas" be?

There were not many options because most substances would be frozen solid at Pluto's temperature range, unable to produce any substantial atmosphere. About the only substances that had the right properties were molecular nitrogen, carbon monoxide (that poison), and molecular oxygen. (Technically, noble gases like argon and neon would also do, but they are so scarce in the solar system that it was doubtful there could be much of them on Pluto.)

A second line of evidence came to the fore in 1992–1993, when a US/French group of planetary astronomers consisting of Alan Stern,

David Weintraub, and Michel Festou set out to check the strange, cold-Pluto result obtained by Altenhoff's team. Seeing the discrepancy between the implications of the occultation data and the IRAS temperature that implied a far higher atmospheric pressure, this team obtained measurements of Pluto's thermal emission using the same telescope in Spain that Altenhoff had used, IRAM, and then verified the result using another radio telescope on Mauna Kea in Hawaii. They found that Altenhoff and colleagues were dead right. Pluto's radio-derived temperature was very close to 40 K.

When the US/French team gave technical talks on their findings and later published them in the journal *Science*, they pointed out that the radio results could square with both the occultation-derived pressure and the radiative equilibrium calculation of temperature *if* nitrogen or carbon monoxide, rather than methane, dominated Pluto's atmosphere. Why? Because the sublimation of nitrogen or carbon monoxide ice at roughly 40 K would produce an atmospheric pressure right smack in the range that the occultation data demanded. Thus, if nitrogen or carbon monoxide were indeed in Pluto's atmosphere as a second gas, it would explain why the occultation scale height was lower than predicted for methane, *and* give a surface pressure in accord with both the occultation measurements and the radio temperatures. It was time to make a careful search to see if these molecules were present in Pluto's spectrum and had been missed.

Something So Familiar

Just such a search was proposed in 1992 by Toby Owen of the University of Hawaii (Figure 4.13), MIT graduate student Leslie Young, and Jim Elliot. Other members of their team included Dale Cruikshank, French astronomers Catherine de Bergh and Bernard Schmitt, Hawaii's Tom Geballe, and Cruikshank's former student Bob Brown, who was then working at JPL (Jet Propulsion Laboratory).

This was not going to be an easy observation. Recall that even a trace of methane is easy to detect, revealing itself as if a bull in a china cabinet. By comparison, nitrogen is a church mouse – it hardly makes itself visible even when it is highly abundant; carbon monoxide is equally difficult to detect spectroscopically. To search for their inconspicuous prey, the team took advantage of a new, far

Fig. 4.13: Tobias (Toby) Owen, who with his colleagues discovered that the dominant ice on Pluto is frozen nitrogen. Photo ca. 1996. (J. Mitton)

more sensitive infrared spectrometer that had just become available on the UK Infrared Telescope (UKIRT) at Mauna Kea. The new spectrometer, called CGAS, was 16 000 times more powerful than the simple photometers Dale Cruikshank, Carl Pilcher, and David Morrison had used back in 1976 to find the methane. It is a good thing too, for Owen and co-workers were going to need all the grasp CGAS could provide.

So, in May of 1993 Owen, Young, and their collaborators ascended Mauna Kea, and pointed the UKIRT telescope with CGAS aboard toward Pluto for a four-night run. With CGAS's incredible capabilities it did not take long. Within days the team knew they had bagged their prey, because the subtle signatures of both frozen nitrogen and frozen carbon monoxide were found in their infrared spectra of Pluto.

Several months of hard work followed Owen and team's initial detection – analyzing the data and then carefully fitting it to laboratory spectra of frozen nitrogen, methane, and carbon monoxide. Using a computer model to vary the concentrations of each of these ices, Owen and his team were able to determine how much of each constituent was needed to best fit their spectra. They found that nitrogen ice comprised about 99.5% of the mix. In contrast, the sur-

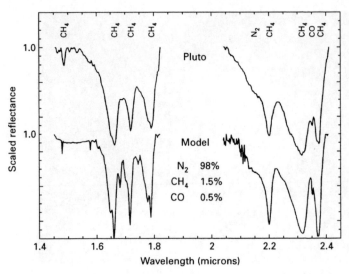

Fig. 4.14: Part of Pluto's infrared spectrum (top trace) compared with a theoretical spectrum calculated on the basis of a 98% nitrogen (N_2), 1.5% methane (CH_4), and 0.5% carbon monoxide (CO) ice. Note how well the theoretical model matches the observation. Some individual features in the spectrum due to the presence of the three frosts are labeled. (Adapted from T. Owen, D. Cruikshank 1997, in *Pluto and Charon*, University of Arizona Press)

face contained only about one carbon monoxide molecule for every 200 nitrogen molecules, and about one methane molecule for every 1000 nitrogen molecules (Figure 4.14). All the while since 1976, methane's dominant spectral signature had lulled astronomers into thinking methane was the dominant ice on Pluto's surface, when in fact it was not!

Next, using vapor pressure calculations that relate the composition of an atmosphere to the amount and temperatures of various ices on the surface, Owen's team used the surface composition measurements to make predictions of Pluto's atmospheric composition. They found that Pluto's atmosphere had to consist primarily of nitrogen, with minor amounts of methane and carbon monoxide.

Leslie Young added an important observational confirmation to this in 1994 when she, Jim Elliot, and several colleagues obtained the very high-resolution spectrum of Pluto's methane absorptions for which researchers had longed for 20 years. With that high-resolution spectrum, Young (then still an MIT graduate student) and her team directly detected methane gas in Pluto's *atmosphere*.[23] For the first time, methane gas had been unambiguously detected above the methane ice. Young's spectrum showed that the methane gas was present only in trace levels in Pluto's atmosphere. This confirmed that the methane signatures detected by Cruikshank and so many others since the mid-1970s had been due, almost entirely, to frozen methane ice, and provided strong corroboration for the case for a nitrogen-dominated atmosphere.

Boom, it was done! Our whole view of Pluto's atmospheric composition had been turned over. The "air" on Pluto, it turned out, was not made of methane. Instead, it is now known to consist primarily of something so familiar – nitrogen, the main constituent of the air you are breathing as you read this page. Of course Pluto's nitrogen is mixed with enough carbon monoxide and methane literally to choke a horse, but there probably are not any horses on Pluto.

Hot Spots on a Cold World

Things were falling into place in rapid succession. But what about the IRAS measurements that gave temperatures of 60 K, far too high to be in agreement with the other data? When the IRAS data were re-examined, they looked fine. When Sykes's data reductions were checked they were bang on. When Sykes's temperature calculations were repeated, they checked. Neither Sykes nor IRAS was in error.

As it now appears, there are *both* warm surface units on Pluto (those 60-K regions to which the IRAS measurements were sensitive) and colder areas (the 40-K regions to which the radio measurements had more sensitivity).

It may be that the 40-K regions are the bright spots and polar caps in the mutual event maps, while the 60-K regions are the darker regions first resolved in those maps. If this view is correct, then the bright regions should be colder in large part because they are where the nitrogen and other volatiles lie, furiously sublimating, and therefore providing an energy sink that cools these parts of Pluto's surface. Likewise, it is thought that the darker regions are warmer in large measure because they are deficient or devoid of much sublimating ice. The fact that the IRAS-derived temperature of 60 K agrees well with the radiative equilibrium temperature calculation we described earlier for Plutonian ground devoid of ice lends strong support to this hypothesis.

In the late 1990s IRAS's successor in space, ISO,[24] measured how Pluto's surface temperature varies as the planet's rotation shows us different amounts of bright and dark terrain. IRAS could not make this measurement because it was not as sensitive as ISO. The ISO data showed that Pluto's infrared brightness does indeed correspond to bright and dark surface regions, which are cold and warm regions at about 40 K and 60 K, respectively. This has important implications for meteorology on Pluto, suggesting the possibility of ferocious winds, complex meteorological dynamics, and just possibly some real weather blowing across the Plutonian landscape.

The Fate of Pluto's Atmosphere: Future Tense

The atmospheric laundry suggested in the late 1980s, both as evidence for Pluto's then putative atmosphere and for Pluto's changing lightcurve between the 1950s and the 1980s, predicted there would come a time after Pluto's perihelion when the planet's increasing distance from the Sun would result in cooling, which would cause the atmosphere to snow out onto the surface and begin to decline in mass.

Sophisticated computer models were constructed to predict how this phenomenon would proceed but models are only as good as the input parameters they are fed. As it turns out, Pluto's atmospheric evolution is highly sensitive to a variety of parameters we do not know well, such as the fine-scale distribution of terrain reflectivities on Pluto, and the relative distribution of the nitrogen, methane, and carbon monoxide snows on Pluto's surface. So models for how Pluto's atmospheric bulk would change over time after perihelion varied from a gentle decline over many decades to a sudden collapse over just a few years. Moreover, the timing of the decline was variously predicted to begin as early as the mid-1990s to as late as perhaps the 2030s.

As late as the early 2000s, it still was not known what we should expect. This was best illustrated by a pair of stellar occultations that occurred in 2002. They were the first Pluto occultations observed since 1988. Because they were predicted well in advance, numerous teams moved portable occultation telescopes to the predicted paths of these events on Earth.[25] They revealed that Pluto's atmosphere had indeed changed dramatically since 1988. But rather than having declined, the atmosphere seems to have increased, doubling its atmospheric pressure! Further, the haze layer or temperature change detected high in Pluto's 1988 atmosphere seems to have vanished. As Nobel laureate Enrico Fermi said so well, "Who ordered that?"

Does the increase (rather than decrease) in Pluto's atmospheric pressure and bulk as it has moved farther from the Sun after its 1989 perihelion mean that atmospheric collapse will not occur? No, perhaps the collapse is simply delayed. Or perhaps the collapse event is complex, resulting in occasional spasms of higher pressure. Or perhaps the atmosphere reached its maximum between 1988 and 2002 at some level above the 2002 pressure, and it is indeed already collapsing. Which of these scenarios is correct (if any!) will require still more data to be collected, for Pluto is again turning out to be more complex than we had imagined. Fortunately, Pluto's track across the sky is moving it against the dense star fields of the galactic center in the constellation Sagittarius. For the decade between roughly 2005 and 2015 many occultations are predicted. Stay tuned.

5
Building a Binary Planet

"The universe is not only stranger than we imagine. It is likely
stranger than we can imagine." – J. S. B. Haldane, 1926

What an intriguing little world Pluto has turned out to be. Far against
the deep, it glides on the ragged edge of our planetary system, accom-
panied by a like-sized satellite, and sporting complex surface markings, a come-and-go atmosphere, and an eccentric orbit that ranges
back and forth by more distance than the entire span from here to
Uranus! How did such a strange and apparently unique world come
to be? How does it fit into the architecture of the solar system?
Was Pluto an oddball or was its discovery the herald of something
commonplace but new? The answers have for a decade been at the
crossroads of the hottest areas of research in planetary science.

A Grand Design

To unravel Pluto's origin we have to begin by exploring the broader
context of the architecture of our solar system and how it all came to
be.

Here is the basic architectural structure of our solar system as it
used to be taught to schoolchildren. It is centered on the Sun, called
Sol, which is immediately surrounded by four inner, rocky planets on
tight-knit orbits. Somewhat farther out, there is the asteroid belt, then
the four giant, low-density gas worlds, and then, finally, lonesome
little Pluto.

But there is more to the house of Sol than meets the schoolchild's
eye. There are also comets, and moons, and planetary rings. There
are little pockets of debris from the era of planetary formation that
orbit in lockstep with giant Jupiter. There are meteorites that have
been blasted off the rocky asteroids and inner planets. And there is
the gargantuan Oort Cloud of comets orbiting the Sun a thousand
times farther away than Pluto itself.

Pluto and Charon, S. Alan Stern and Jacqueline Mitton
Copyright © 2005 WILEY-VCH Verlag GmbH & Co. KGaA, Weinheim
ISBN: 3-527-40556-9

Is not it amazing that this inanimate collection of objects we call our solar system has any design at all? After all, why are not the giant planets interspersed among the rocky ones? Why are the rocky planets all close to the Sun? Why is there no large planet in the gap between Mars and Jupiter? Why are the solid inner worlds so much smaller than the gas giants lying farther out? Why are most planetary moons ten thousand times or more smaller than their parent planet, and yet Luna is a quarter the size of the Earth? Why are the giant planets topped with massive atmospheres? And where did all the material, the almost five hundred Earth masses that are now contained in the planets, come from anyway?

At the dawn of the twentieth century, one could pose such questions, but the answers remained hidden behind a veil of ignorance. Today, at the dawn of the twenty-first century, we are lucky to enjoy the fruits of decades of intensive study using telescopes, computer simulations, probes to every planet save Pluto, and, importantly, half-a-dozen human explorations of our ancient neighbor, Luna. As a result, it is now possible to recount with some confidence how the solar system came to be as it is.

The journey towards obtaining a basic, but still incomplete understanding of our solar system's origin and evolution has been a tortured and expensive one. More than a thousand humans from continents around the world have invested their careers in it. It has cost billions of dollars, and several lives, and it is not yet complete.

There is little wonder why this journey is still incomplete: the story is more complicated, more subtle, and more convoluted than many expected it would be. Indeed, another century hence, our grandchildren, and their children too, will no doubt still be unraveling the story. But the thread of exploration that spans generations of astronomical explorers is itself a part of the lure of this quest to understand our origins.

Before

Comedian Steve Martin once said that he knew the secret to getting a million dollars and never paying taxes: "It's easy. First, you get a million dollars ... " The secret to forming the solar system is not so different: First, you get a universe.

But our story begins two-thirds of the way through the history of the universe. At that time, our galaxy the Milky Way was about 4.6 billion years younger, but was mature enough to appear much as it does today. About 26 000 light years from the galactic center, out in the spiral arms where the Sun quietly orbits, a massive interstellar cloud of dust and gas was collapsing under its own weight, possibly triggered by the passing shock wave of a nearby, exploding star. As the collapse proceeded, the cloud fragmented into clumps, many of which would become individual stars, complete with their own planetary systems. The collapsing cloud was composed primarily of hydrogen and helium; in fact, just two or three percent of the cloud mass consisted of anything else.

As the cloud fragment with the Sun's name on it – the "proto-solar nebula" – collapsed, it heated up. This heating was due to the compression of the gas that the collapse was all about. The collapse also enormously amplified the cloud's spin. Like a skater pulling in her arms during a pirouette, the cloud spun round faster and faster as it shrank. And shrink it did, for the proto-solar nebula began perhaps a light year or two across, and ultimately shrank by a factor of 1000 or so to form the solar system. Quickly, which in our context means in a lot less than a million years, the solar nebula formed an ever more intensely smoldering protostar at its center with a flattened disk of dust and gas orbiting around it. The disk probably extended out hundreds of astronomical units[26] from the infant Sun (Figures 5.1 and 5.2).

At the same time, in the little corner of galactic space within a few tens of light years of the young Sun and its disk, hundreds of similar young stars and their disks were also forming from their own fragments of the same giant galactic cloud. Most of the stars that formed in the same cluster as the Sun were less massive than Sol, but some were more massive, and a fair fraction were more or less identical to the Sun. Some stars formed alone, as did the Sun, but others formed in pairs (binaries), triples, or – very rarely – even larger groupings.

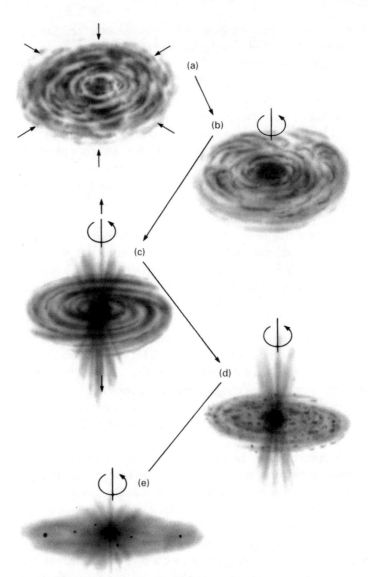

Fig. 5.1: The main stages in the formation of the Sun and the solar system are schematically illustrated. (a) The cloud of interstellar gas from which the Sun and the planetary system formed begins to collapse. (b) As the cloud collapses, its rotation speeds up and it takes on a flatter shape. The center becomes hotter and denser. (c) The proto-Sun forms, surrounded by a disk of gas and dust. The young Sun blasts jets of material along the direction of its rotation axis, while more gas and dust falls into the disk. (d) Material in the disk clumps together to form planetesimals. Accretion of larger bodies takes place, along with destructive high-speed collisions. (e) The solar wind sweeps away any remaining gas as the major planets we know today emerge as the survivors.

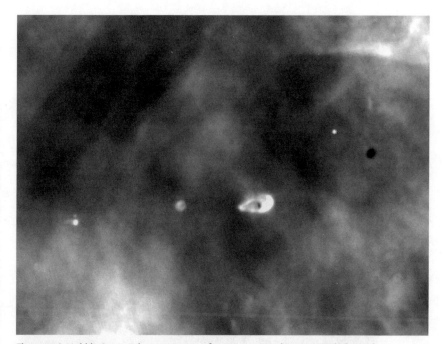

Fig. 5.2: A Hubble Space Telescope image of young stars in the Orion Nebula, each surrounded by a disk of gas and dust that may in time evolve into a planetary system. (C. R. O'Dell, Rice University, and NASA)

Returning to our own star's spinning disk of material, several interesting things were taking place at this early stage. For one thing, drag within the disk was causing dust and gas to spiral inward. Some of it fell onto the Sun, adding mass and affecting its early spin rate. Some of the material was funneled way in opposite directions along its polar spin axis in two great exhaust plumes that astronomers call "bi-polar outflows." These plumes reduced the disk's mass and also helped to cool it (Figure 5.3). Very importantly for our story, the dust grains within the disk grew and grew. Exactly how the grains grew is not fully understood yet, but observations of star forming regions in space support the case that, somehow, through low-velocity collisions or larger scale gravitational collapse, or maybe even due to electrostatic effects, dirt grains, ice grains, snowballs, boulders, and, ultimately, larger objects called planetesimals built up.

The process that began with a cloud collapsing to a disk, and ended with the formation of planetesimals, is a great concentrator of heavy

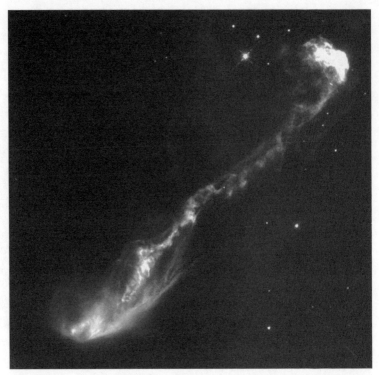

Fig. 5.3: A jet of gas 1.5 light years long bursting out of the dark cloud of dust and gas at the lower left. This cloud conceals a newly forming star. The jet is known as HH47. The image was taken with the Hubble Space Telescope. (J. Morse, STScI, and NASA)

elements.[27] As a result of this concentrator effect, the planetesimals that formed consisted primarily of heavy elements like carbon, nitrogen, oxygen, and sulfur. The planetesimals, we know from studies of asteroids, comets, and meteorites, contained only trace amounts of helium, and perhaps 10% hydrogen (primarily bound up in water ice or hydrocarbon molecules). The reason that the planetesimals were so enriched with heavier elements over hydrogen and helium is straightforward: neither hydrogen nor helium can condense easily or become trapped in great amounts within the solid particles in the disk – for it is simply too warm for hydrogen and helium to condense there as solid grains. As a result, most of the hydrogen and helium remained in gas form, while much of everything else, which condenses into solid ices and minerals at significantly higher temperatures, ended up within the planetesimals.

Building Planets from Planetesimals

The planetesimals, formed as described in the previous section, were the solid building blocks of the planets. Close to the Sun, where temperatures were too high for water ice to remain condensed (as it was in the in-falling material of the proto-solar nebula), the planetesimals consisted primarily of rock. Farther out, probably somewhere near the place where Jupiter is today, conditions were cold enough to allow water ice to remain frozen. As a result, the planetesimals that formed at and beyond this distance were probably much like the comets we see today – fragile bergs of ice and dust.

As time passed, the planetesimals grew. To develop a picture of how this growth took place, imagine the view from the surface of a planetesimal. All around would be the cloud of the nebular disk. Close to the Sun there might actually be enough light to see, but if it is farther out, the dense gas and dust between it and the Sun is thick enough to cast a darker pall than any bad weather day does beneath the blackest clouds of Earth.

One of the most important processes acting in the disk is the aerodynamic drag on solid objects created by the nebular gas. The root cause of this drag is that the gas molecules in the disk travel slightly slower than the solid particles.[28] The larger the object, the less it is affected by the gas. As a result, boulders and planetesimals are not as strongly influenced by the aerodynamic effect of the nebular gas flowing over it. However, smaller objects – like dust grains, snowflakes, pebbles (because of their larger area-to-mass ratio) – feel the drag to a greater extent, and spiral sunward faster than large objects.

Aboard a large planetesimal, one would (flashlight in hand) observe a continual accretion of pebbles and rocks, along with less frequent collisions from boulders, and house- and even hill-sized entrants spiraling sunward and occasionally running smack amidships. Even more rarely, another planetesimal of similar size would come looming out of the murk to collide.

In terms of planet building, collisions are good. For this is the way that objects grow. Fortunately, the same gas drag that causes objects to spiral inward and cross paths also keeps their characteristic collision speeds fairly low – between a few millimeters and a few meters per second – so collisions are gentle as long as there the gas remains, thereby promoting growth.

Of course, not all objects grow at the same rate. Bigger objects grow more rapidly, because they are larger, and therefore simply run into material at a faster rate than the little guys do. Also, planetesimals grow more quickly in the inner solar system than farther out. Why? In part because as one moves farther and farther out, the density of material drops, making collisions progressively more infrequent, and in part because the orbital speeds are higher, which also increases the rate of collisions. As a result, the time it took planetesimals to double in mass in the terrestrial planets region may have been ten times less than in the region where the giant planets are.

Once a planetesimal reaches a size of about a hundred kilometers in diameter, something else occurs to help promote growth: it becomes massive enough that its own gravity begins to play a role, focusing objects coming near it onto collision courses. The cognoscenti call this an enhancement of collisional cross section due to gravitational focusing.

Jargon aside, what matters is that, once a planetesimal becomes large enough to generate appreciable gravitational focusing, the accretion process can cause it to experience "run away" growth. Consequently, it grows at faster and faster rates as the mass it accretes generates more and more focusing, which in turn accelerates its accretion further.

If you are starting to get the feeling that the early solar system looked a lot like a kind of 1920s Keystone Cops movie, with both hits and near misses occurring all the time, then you are getting the right picture.

Once planetesimal seeds have grown to the point where they are about 1% of the mass of the Earth (i.e., have a diameter somewhere near 500 to 1000 kilometers), they become large enough to significantly perturb each other's orbits each time they pass one another. Such objects are often referred to as proto-planets. This mutual jostling about of the proto-planets prevents them from becoming isolated in narrow little "feeding zones" that would otherwise eventually empty as each planetesimal ate up everything within its limited grasp.

Instead, these wolves forage about, ranging over tens of millions of kilometers (or miles, if you prefer).

In the inner solar system, orbits are so tightly bound to the Sun that the growing planets can move things around, but few objects are so disturbed that they are ever actually ejected from the region. Thus, virtually all of the planetesimals in the inner region – where Mercury, Venus, Earth, and Mars formed – had to end up either in these planets, or in the Sun.

In the outer solar system, however, the combination of less tightly bound orbits and the larger size and gravity of the planets that formed there, allowed a lot of the planetesimal supply to be ejected from the region altogether – much of it to interstellar space.

The growing planets in the outer solar system could do more than just focus other planetesimals in. They were also massive enough to be able to capture large amounts of gas from the disk. The technical name for this process is a "hydrodynamic accretion." The best mental picture for it is probably a kind of whirlpool in which gas from the solar nebula is funneled onto the growing planetary cores of the outer solar system at astounding rates.

Why did this process work in the outer solar system but not the inner solar system? A key factor is that the colder conditions in the outer solar system reduced the critical mass a planet had to reach to initiate hydrodynamic accretion. Thus, whereas a planet at the Earth's distance from the Sun would have had to reach a mass of 50 or more Earth masses to initiate the rapid, hydrodynamic accretion of gas, a planet at 5 AU from the Sun, where Jupiter is now, only had to reach 10–30 Earth masses. Of equal importance, the outer solar system contained enough material for planets with masses this large to form. The inner solar system, by contrast, did not contain enough material to make planets large enough to initiate hydrodynamic accretion.

Once hydrodynamic gas accretion begins, it proceeds very rapidly. In perhaps only hundreds of thousands of years – no time at all astronomically speaking – the accretion of gas adds on many dozens of times the mass of Earth onto a growing core like Jupiter's. As a result, Jupiter's core, with a mass of perhaps 20 Earth masses, captured 300 Earth masses of gas in less time than it took the Earth itself to form! Likewise, Saturn's core captured 70 or 80 Earth masses, and the cores of Uranus and Neptune each managed to gather up a few

Earth masses of gas. These four worlds would have kept on growing, too, if something had not intervened.

Winds of Change

The environment in which the planets were growing began to transform radically when the Sun shed its birth cocoon and disk during an intense phase of early stellar evolution called the T Tauri stage. The T Tauri stage got its name from the star in which it was discovered, and observations since the 1970s have shown that just about every solar-type star undergoes the T Tauri transformation by the time it is 10 million years old.

During this phase a fierce stellar wind blows, typically at ten or more kilometers per second. And with this hurricane of hurricanes, virtually all of the hydrogen and helium gas in the Sun's disk was blown back into interstellar space. In stripping the Sun of its disk, the T Tauri stage brought light where there had been only darkness. It left behind little gas, but all of the swarming numbers of planetesimals and proto-planets that had grown large enough to resist its wind.

You can probably already see in this story something that it took decades for researchers to appreciate: timing played a crucial role in determining the final architecture of the solar system.

For example, one of the critical "races" that took place (unbeknownst to its inanimate participants) was between the internal clock in the Sun, racing forward to its T Tauri stage, and the growth of the giant planets. If the Sun had evolved faster, or the giant planets a little slower, then the disk of gas from which the giant planets grew so fat might have been blown away before hydrodynamic accretion began. Had that happened, Jupiter and Saturn would probably have remained stillborn giants, perhaps only 5–15 Earth masses in scale. But in our solar system, Jupiter and Saturn were able to accrete tremendous amounts of gas before the T Tauri wind stripped the gas away. Interestingly, it appears that farther out in the solar system, Uranus and Neptune were growing at a slower rate. These two apparently got to the hydrodynamic accretion stage so late that they had only just begun accreting gas when the Sun went into its T Tauri stage, explaining their lower masses.

Another critical race that took place was the competition between the growth of a planet just beyond Mars, and the growth of Jupiter. Jupiter, it appears, won this race by reaching the hydrodynamic gorging stage well before a massive planet could form out of the rocky material beyond Mars.[29] As a result, Jupiter's newly garnered giant mass exerted a strong gravitational influence on the region where the asteroid belt is today. As a consequence, collisions that should have been gentle enough to promote growth instead resulted in shattering. In effect, Jupiter's massive presence had the effect of terminating the growth-nurturing environment in the region where the asteroid belt is today. The result is a region devoid of a substantial planet containing only the scattered debris left over from the era of shattering collisions that Jupiter induced.

There were other races too, but what all this means is that the detailed character of the solar system is in part due to random chance – such as "who" won which races. As we next explain, other random factors were important as well. As we discover the structure of other planetary systems, we are finding that the same processes acted differently in different locales, introducing a wide variety of planetary system architectures around the Galaxy.

Mother Violence

In the late 1970s, at about the same time that the consequences of the T Tauri stage of stellar evolution were becoming appreciated, researchers typified by George Wetherill of the Carnegie Institution and Bill Ward, then of Cal Tech's Jet Propulsion Lab, discovered something of equally fundamental consequence for planet formation.

Wetherill and Ward found that making planets from planetesimals was not a problem. In fact, computer simulations constructed by Wetherill made lots of little planets the size of Earth's Moon and of Mars. In some of Wetherill's computer simulations, hundreds of lunar-sized and dozens of Mars-sized objects formed in the region between Mercury's orbit and the main asteroid belt. These objects, Wetherill found, were not the final set of planets that were to form, but a crowd of intermediate, transitory "planetary embryos."

Why were these embryos transitory? Simply put, when you crowd a couple of hundred small planets into the inner solar system and let them run about, they eventually start to collide and combine, which reduces their numbers. Ultimately, collisions result in a situation where only a small number of far larger planets remain, their orbits separated by enough space to isolate themselves from further collisions. One such survivor was Earth.

Wetherill's computer simulations beautifully illustrate that, within a few tens of millions of years (i.e., about 1% or 2% of the present age of the solar system), the inner solar system "cleans itself up" through this cascade of collisional combination. Different simulation runs produce different variations on the end result, but most terminate with an inner solar system containing three to six planets with typical masses ranging from about a tenth to perhaps a few times Earth's mass.

Later computer simulations by others convincingly showed that cataclysmic collisions between planet-scale bodies were common in the final stage of planetary formation. Some of these impacts were so massive and so violent that they could melt a planet's whole surface and at the same time reorient its spin axis. It is thought that just such a collision stripped away much of the original mantle of Mercury, leaving behind the dense core and overlying crust that we see today. This great violence probably also caused numerous collisions that vaporized oceans or even melted the surface and deep interior of the Earth many times. In doing so, life may have lost its footing on numerous occasions as Earth's molten surface was repeatedly re-sterilized, but that is another story.

A Misfit's Role

The tale of solar system formation just told was woven from almost four decades of increasingly sophisticated computer modeling and is buttressed by the physical evidence ranging from the ages and compositions of meteorites and lunar samples to telescopic observations of star formation across the Galaxy and spacecraft visits to every major type of body in the solar system out to Neptune.

The result is a tidy, if still sketch-work, consensus about the basic path that led from the solar nebula to the rocky inner planets, an asteroid belt, and then a set of gas giants of decreasing size with distance from the Sun. The scenario we described also does a nice job of fitting in the comets – icy planetesimals, really – most of which were catapulted from the planetary region by gravitational action on the part of the giant planets and now reside in distant Oort Cloud orbits or as flotsam among the stars. If there is something that sticks out like a sore thumb in this scenario, it is the question of how Pluto came to be.

Many people wonder, even as schoolchildren, why Pluto does not seem to fit into the pattern of the solar system. Why is it in that crazy orbit? Why is it out there ten thousand times smaller than its giant planet neighbors? Why does it have a satellite half its own size when no other planet does? After all, planetary formation is such a messy process. Why would the solar system create just one, little, lonely pipsqueak planet just beyond the gas giants? Astronomers wondered too. But for decades, they did not have an answer. Textbooks often called Pluto a planetary "misfit."

As it turned out, the question of Pluto's place and origin in the solar system was unexpectedly deep and profound and, in coming to understand this story, a paradigm shift of revolutionary proportions about the content of the distant outer solar system would have to take place.

The Faint Smell of a Fish

Almost as soon as Pluto was discovered, astronomers began to realize that this new planet was different from the others. The key attributes that set Pluto apart from its planetary kin were known as early as the 1930s: its cockeyed eccentricity and inclination, that Pluto's orbit crossed Neptune's, and the then-emerging evidence that Pluto was much smaller than Neptune.

In 1936 astronomer Ray Lyttleton collected these facts together in an influential research paper published in the British research journal *Monthly Notices of the Royal Astronomical Society*. There Lyttleton outlined several possible scenarios for Pluto's origin. The one that caught on argued that Pluto was an escaped satellite of Neptune. And how did it escape? Lyttleton conjectured it had suffered a close approach to another of Neptune's satellites, largish Triton, which unlike any other major satellite of the planets orbits its parent world on a reverse course. In the ancient gravitational exchange that Lyttleton imagined, Pluto had been ejected and Triton's orbit had been reversed. Thus, Lyttleton's scenario offered a tidy, economical explanation for why Pluto is so small (it used to be a satellite), why Triton orbits Neptune backwards (it had been reversed by an encounter with Pluto), why Pluto's orbit appears so rakishly disturbed (it was ejected from orbit about Neptune), and why Pluto apparently crossed Neptune's orbit.

Lyttleton's scenario enjoyed a broad public exposure, and gained a wide, tacit approval from astronomy textbook authors. Despite the fact that evidence mounted against it over the decades, as late as 1995, when Lyttleton died, most astronomy textbooks still offered his 1936 scenario as the standard explanation of Pluto's origin. Unfortunately, Lyttleton was wrong.

What was wrong with Lyttleton's scenario? The first problem emerged from exquisitely detailed simulations of the planetary orbits (Figure 5.4). This kind of calculation is computationally intensive because accuracy demands recalculating the positions of each planet to within a few meters tens of billions of times, and each planet's position affects all of the others. When computers became fast enough in the mid-1960s and early 1970s to simulate accurately the evolution of planetary orbits for millions of years, Pluto's strange orbit received close scrutiny. What researchers making these kinds of computations found is that, despite the simple appearance of Pluto's orbit crossing Neptune's, the two never *actually* come close to one another at all. This is because Pluto's orbit and Neptune's are locked in a precisely repeating pattern called a resonance that prevents close approaches. At the heart of this resonance is the amazing circumstance that Neptune's orbital period is *exactly* two-third's that of Pluto's. This repeatability means that, every time Neptune makes three laps around the Sun, Pluto has made exactly two. This synchronicity (which astronomers refer to as Pluto's "3:2" resonance with Neptune), and the orientation of Neptune's orbit relative to Pluto's, ensures the two always remain far separated. Whenever Pluto crosses inside Neptune's orbit (as, for example, it did from 1979 to 1999), Neptune lies about a quarter of the way around the Sun (from Pluto), and Pluto is at the northern extension of its inclined orbit (Figures 5.5 and 5.6). Although it is not known how this came to be, it is a fact. And it means that even over billions of years, Neptune and Pluto cannot come closer to one another than about 2.2 billion kilometers (1.5 bil-

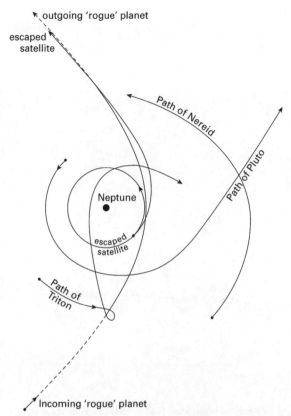

outgoing 'rogue' planet

escaped satellite

Path of Nereid

Path of Pluto

Neptune

escaped satellite

Path of Triton

Incoming 'rogue' planet

Fig. 5.4: After Raymond Lyttleton first proposed in the 1930s that Pluto may have originated as a satellite of Neptune, a number of different attempts were made to devise scenarios in which Pluto might have escaped from Neptune. The particular scenario shown here requires a "rogue planet," of about 3 Earth masses, to pass close to Neptune and eject Pluto. In one deft action the rogue is shown disrupting the paths of Triton and Nereid to their unusual present-day orbits, and ejecting both Pluto and Charon completely out of their Neptunian orbits. Such a scenario is now regarded as virtually impossible, not least because the relation between the orbital periods of Neptune and Pluto ensures that the two never even remotely approach each other, and therefore that Pluto could not have originated in orbit about Neptune. (Adapted from R. S. Harrington, T. Van Flandern 1979, *Icarus* 39)

lion miles). What wonderful machinery! No Swiss watch could be so fine. This alone, this orbital quarantine of Neptune from Pluto, deals a fatal blow to Lyttleton's scenario. Pluto could not have originated in orbit about a planet it cannot approach.

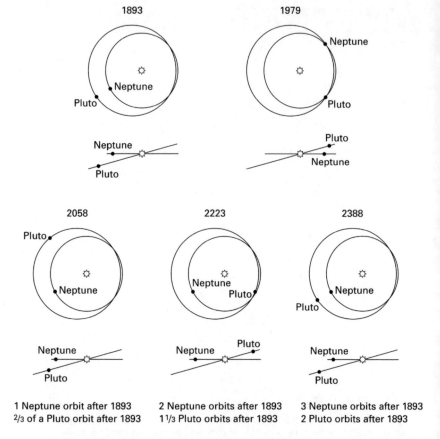

1 Neptune orbit after 1893   2 Neptune orbits after 1893   3 Neptune orbits after 1893
2/3 of a Pluto orbit after 1893   1 1/3 Pluto orbits after 1893   2 Pluto orbits after 1893

Fig. 5.5: Though Pluto's orbit crosses that of Neptune, the two planets are never in danger of colliding. This is because they are locked into orbits with stable periods in the ratio 3:2, which, along with the orientation of the two orbits, ensures that the two stay well apart. In fact, Pluto gets nearer to Uranus than it ever gets to Neptune. Even so, it never approaches Uranus closer than a billion miles. Upper left: Neptune and Pluto are only ever lined up on the same side of the Sun (at conjunction) when Pluto is near aphelion, as in 1893, for example. Upper right: When Pluto crosses the point where it becomes nearer the Sun than Neptune, the two planets are well separated as seen either face on or side on to Neptune's orbit. Bottom row (left to right): Pluto and Neptune reach conjunction again only after Neptune has completed exactly three orbits to Pluto's two.

And as our knowledge of Pluto, Neptune, and Triton grew in the 1980s, Lyttleton's scenario suffered a second, equally fatal wound. Enter Bill McKinnon, a Caltech-minted PhD interested in planetary origins and interiors (Figure 5.7). McKinnon is one of those rare,

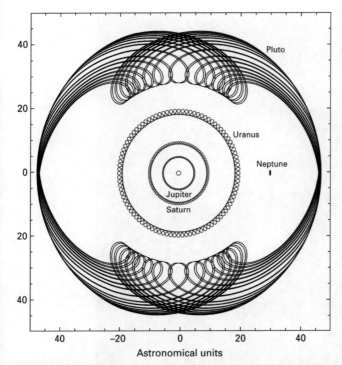

Fig. 5.6: The way Pluto keeps its distance from Neptune is neatly depicted in this diagram, which shows the motion of Jupiter, Saturn, Uranus, and Pluto over a 40 000-year period from the point of view of a stationary Neptune. In this reference frame, two orbits of Pluto trace out a complete loop in just under 500 years. The figure illustrates how the relative orientation of the orbits of Pluto and Neptune slowly goes through a 20 000-year cycle (a phenomenon called libration), but never results in a close approach of the two planets. (Adapted from R. Malhotra, J. Williams 1997, in *Pluto and Charon*, University of Arizona Press)

brilliant mavericks, who are deeply respected both for creativity and for thoroughness. What McKinnon did was to show that Pluto's mass is both too small to have reversed Triton's orbit and too large to have been ejected by Triton. If the orbital simulations had shot Lyttleton's hypothesis dead, McKinnon's calculations sealed the coffin.

As weaknesses in Lyttleton's hypothesis emerged, astronomers turned to other potential origin scenarios for Pluto. "Perhaps Pluto was an escaped asteroid." No – how would an asteroid get all the way out where Pluto is and then get locked in that strange resonance with Neptune? "Ah, maybe it's a failed giant planet core that never accreted gas because the Sun's T Tauri clock ran faster than Pluto's

Fig. 5.7: William McKinnon in about 1996. Among other things, he showed that Pluto could not have been ejected by Triton from orbit around Neptune. (J. Mitton)

growth." Probably not – Pluto is far too small to be a giant planet core. "Oh, maybe Pluto *was* a giant planet that somehow lost its atmosphere to reveal the denuded seed inside." But no process that could do this is known (and why did not Uranus and Neptune suffer similar fates?). Reading the developing scientific literature of the day, you can almost see the sweat as researchers struggled to find some explanation for the little misfit planet. There was a faint smell of something fishy in the air about Pluto and its curious situation but no could quite put their finger on it.

With the advantage of hindsight, it is easy to see that the old ways of thinking about this problem were simply too limited in their imagination, too restricted by inadequate data, and too rooted in the need to explain Pluto as an oddity. To explain Pluto's origin, something very different was required, something wonderfully new, something almost wholly unexpected. It would be something that helped to open a revolution in our whole concept of the architecture of the outer solar system. Strangely enough, the first clue came from a place very far away from Pluto indeed – from Earth's Moon.

Big Bangs

Luna had been a subject of serious scientific inquiry since Galileo first turned his telescope towards it at the dawning of the seventeenth century. And perhaps the most fundamental scientific question about the Moon posed over the centuries between Galileo's time and ours has been the issue of how it came to be. By the time the Apollo program was under way in the late 1960s, three contending theories had emerged, each with its own proponents, but dispassionate observers of the debate agreed that none of these scenarios fit the lock with a proper key.

One of the three competing scenarios was called the fission theory (a kind of ultimate "windup and pitch"), in which the Moon was thought to have been shed – literally flung off – from the young, rapidly spinning Earth. However, this fission scenario suffered from the fact that repeated calculations had shown that there was no natural way to get the early Earth spinning fast enough to split itself into the Earth and Moon.

A second idea was that the Moon had formed as a planet in orbit around the Sun but had somehow been captured into Earth orbit. Unfortunately, computer simulations of capture found the process very unlikely to occur and, when it did, it often resulted in a collision between the Earth and Moon that destroyed the Moon.

A third idea was that the Moon had accreted in orbit around the Earth as the Earth itself formed. But calculations could not explain the large mass and orbital momentum of the Moon if it formed this way.

All of these ideas had emerged by the middle of the twentieth century but there was no clear favorite. By the 1960s, most planetary scientists and astronomers expected that the rich scientific harvest of samples and geophysical data that Apollo was expected to bring would shed enough light to make clear which of the three theories was correct, or correct enough to be patched up.

However, the rocks and other evidence brought back from the Apollo lunar expeditions set the whole study of lunar origins on its ear. After Apollo, none of the theories looked viable! Why? In large part, this was because the proportions of different chemical elements and their isotopes in lunar samples resembled the Earth's deep mantle, rather than the material closer to the surface.

How, researchers asked, could this have been the case if the Moon had spun off the Earth, as in the fission hypothesis, or if the Moon had formed either in solar orbit as its own planet or in orbit around the Earth? Just as it was with Pluto, the old ideas about lunar formation had failed so completely that a wholly new idea was required. Amazingly, the idea that came to triumph sprang forth almost simultaneously in two different places.

Don Davis and Bill Hartmann from the Planetary Science Institute in Tucson, Arizona, hit on it first, in 1975. Al Cameron and his colleague Bill Ward at Harvard thought of it almost simultaneously.

The Moon, these groups offered, had been dug out of the young Earth in a cataclysmic collision between the early Earth and one of the Mars-sized planetary embryos that existed during the final stages of inner planet accretion. This theory, which came to be known as the "giant impact hypothesis," fit beautifully into the then just-emerging view of late-stage planetary accretion, in which dozens, if not hundreds, of lunar-sized and larger planets were competing with one another for space and mass, and in the process of doing so, suffering mutual collisions of gargantuan scale. The idea that massive collisions contributed to the formation of the planets is widely accepted now but it was well ahead of its time in the mid-1970s, when most researchers still thought planets grew only by the slow sweeping up of tiny planetesimals and small debris.

Once Hartmann and Davis, and Cameron and Ward, had described their fresh idea, rapid advances began to be made. These two teams, and then many others, began testing the giant impact hypothesis against computer models, chemical evidence from Apollo samples, and what was known about the Earth's mantle.

Computer simulations of the giant impact found it possible to put a lunar mass in Earth orbit, and to do so with the right orbital momentum and chemical signatures to match the real Moon – something

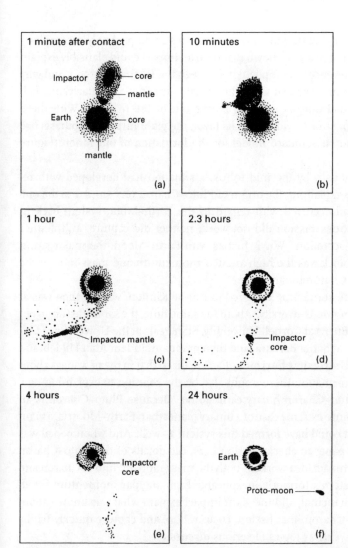

Fig. 5.8: A giant impact in action. The possible sequence of events if the newly formed Earth were struck by a large body (about 15–30% of Earth's mass) in order to form the Moon. Both Earth and the impactor have metal cores overlain by mantles of silicate rock. After the collision, the impactor is broken up (c), but the debris clumps together, largely through the action of gravity. The iron core of the impactor separates out and falls into the Earth (d and e). Some of the debris from the impactor's mantle accretes in orbit around the Earth to form the Moon (f). (Based on computer simulations by A. G. W. Cameron and W. Benz)

none of the earlier lunar origin theories could plausibly accomplish. Further, it was also shown that such an impact could plausibly explain the difference in composition between the Earth and Moon, and why the Moon looked so much like Earth's mantle: the cataclysmic collision had dug its building blocks out of the mantle! With these kinds of strong evidence in its favor, the giant impact hypothesis has become *the* standard model for the formation of our Moon (Figure 5.8).

Curiously, by the mid-1980s, a situation had developed with regard to explaining the origin of Pluto–Charon very similar to the one that had occurred with regard to the Earth–Moon system back in the 1960s. Fission did not work; neither did capture, and neither did co-accretion. Why? Just as with Earth–Moon, the major stumbling block was the high angular momentum and mass ratio of the Pluto–Charon binary.

Then, something occurred to Bill McKinnon, who by then was at Washington University. About the same time, the same idea occurred to Joe Burns at Cornell University, Stan Peale at the University of California at Santa Barbara, and then graduate student Alan Hildebrand at the University of Arizona. Each realized that in the new ideas about lunar formation there might also be the solution to explaining how the Pluto–Charon binary came to be. Because Pluto–Charon is an even more extreme case of a binary planet than Earth–Moon is, a giant impact could have formed this system as well. And when pencil was put to paper to check the numbers, the details of the Pluto–Charon giant impact idea worked: both the mass ratio of Pluto to Charon and the system's formerly inexplicable high angular momentum fitted! So convincingly did the giant impact explain what was known about Pluto–Charon, that fission, co-accretion, and capture quickly fell by the wayside as topics of serious discussion.

Today no clear rival to a giant impact exists for the formation of the Pluto–Charon binary. In fact, recent, detailed simulations by Robin Canup of the Southwest Research Institute, one of the foremost lunar formation theorists, has convincingly shown that the giant impact hypothesis nicely fits all of the known attributes of the Pluto–Charon binary.

Lost Flock

The giant impact model indicates strongly that the Pluto–Charon binary formed in a scenario rather like this. First, Pluto must have formed in the ancient outer solar system. Why it was so small compared to Neptune and the other giants is not entirely clear but it probably had to do with Pluto's growth being cut short in the ancient past. That Pluto was a creature of the icy outer solar system was fairly certain: after all, its surface ices were too volatile to have persisted in the inner warm regions. Even more importantly, Pluto's average density proved it contained a good deal of water ice, which, of course, the rocky inner planets and asteroids do not. Then, sometime after Pluto formed, another large object, with a mass of 30–50% of Pluto's mass, struck it. The impactor object could not have been much smaller than 1/3 of Pluto's mass (and thus perhaps 2/3 its diameter depending on its density), or else it would not have packed the punch to create the binary. In the aftermath of the blow, enough material was stripped from Pluto's surface and into orbit to create Charon. From that material, Charon accreted in orbit around Pluto.

This is a nice scenario for the formation of the binary itself, but it begs several important questions. For example, where did the Charon-forming impactor that collided with Pluto come from? And how did the impact just happen to be the right kind to create a binary? A gentle collision would instead have created a merged world, and a sufficiently violent one would have shattered both Pluto and the intruder to smithereens. So too, why would the solar system make just one, highly unique, tiny binary on its outer border? After all, the outer solar system looked pretty empty – where could the impactor have come from, and what were the odds it would have struck little Pluto? One of us (Alan Stern, then a postdoctoral worker at the University of Colorado) was the first to examine this question.

The natural way to approach this key question is to estimate the amount of time it would take Pluto and the impactor body, in similar orbits, to run into one another. Timescale questions like this can be answered in a rough but robust way by a method called a particle in a box (PIB) calculation. The PIB theory was originally developed in the nineteenth century for the molecular theory of gases, but it can be applied in many other situations as well, and it is a common tool used by astronomers to estimate timescales.

The PIB equation for calculating the timescale for object A to be hit by object B only requires knowing three things about the situation: (a) how big the target (object A) is, (b) what the concentration of projectiles B is, and (c) how fast the projectiles are moving about relative to the target. Given these three numbers and a hand calculator, one can quickly find the approximate time it would take for such a collision to become likely. Let us see how the factors that go into this calculation are arrived at.

At first sight it might seem that the target size for our situation would just be the size of Pluto. However, it is more realistic to recognize that a collision occurs whenever the target and the projectile touch. Therefore, if the projectile is a good fraction of the target's size (as a Charon-forming giant impactor must be), then the target size is best represented by the sum of Pluto and the impactor. To make properly the calculation, one should also take account of Pluto's gravitational focusing.

To estimate the concentration of projectiles (i.e., how many of them were packed per cubic kilometer), the simplest and most conservative thing to do is to assume there was just one impactor in the whole volume of space that Pluto's orbit draws out over time. That is about one object per 30 billion, billion, billion cubic kilometers.

To estimate the speed at which the projectile and target stumble around, blindly searching for one another, one needs to know their typical orbital eccentricities and inclinations. Since it is impossible to know what the eccentricity of the giant impactor's orbit *actually* was, one can instead solve the PIB equation for the whole range of plausible encounter speeds.

Plugging in the numbers gave an astonishing result. Regardless of whether the impactor and Pluto had a low or high relative speed, it is hard to make them collide in a reasonable amount of time. The space is just too big for them to "find" one another very quickly.[30]

The specific finding of the PIB calculation was this: if Pluto and a proto-Charon were let go to wander about in deep outer solar system in orbits like Pluto's, they would take between 10 000 and 10 000 000 times the age of the solar system to collide![31] Put another way, in the age of the solar system, the odds of Pluto being hit by a single giant impactor orbiting in the same region of the solar system were between 1 in 10 000 and 1 in 10 000 000. So, how could this be reconciled with the convincing case that a giant impact is needed to create the Pluto–Charon binary?

The obvious answer was to put more than just one impactor in the mix. With more projectiles running about, Pluto would get hit more often. As it turns out, to make the collision likely in less than the age of the solar system, the mathematics dictate that between 300 and 3000 Charon-forming impactors had to be roaming about in the early days of the solar system.

The implications of this were as follows. On the one hand, if the early outer solar system contained only Pluto and the Charon impactor, then our best hope for explaining this binary planet, the giant impact hypothesis, had a ridiculously low probability of having actually occurred. On the other hand, if the impact did occur, then it implied that Pluto was not born a lonely little footnote to planetary formation, but is instead the telltale survivor of a formerly large population of similarly sized objects that were present in the ancient outer solar system. Which was it to be?

6

Ice Fields and Ice Dwarfs

"Where have all the flowers gone, long time passing?"
– Ecclesiastes

If Pluto had been the only smoking gun to suggest the presence of large numbers of now-missing objects in the ancient outer solar system, the story might have stopped right there. After all, the formation of the outer solar system was so rough and tumble that some very unlikely things might well have happened. So, although the Pluto–Charon binary could be a fluke, it is a bit much to accept that it is a freak and the only survivor of the birthing mêlée of the outer solar system. Taking a lesson from paleontology, when a strange new fossil is discovered, one has to assume that the find is a relic of a formerly widespread species, as opposed to a mutant that just happened to survive the ages. By analogy to Pluto–Charon, this might mean that more Plutos or even more Pluto–Charons might have once existed. But is there actually evidence supporting the case for a vast population of ancient ice dwarf planets?

The first such evidence came from studies of how planets got their tilts. By the late 1980s it was pretty well accepted that the tilts of the inner planets came from the crashing and combining together of George Wetherill's lunar-to-Mars-sized bodies into the growing rock worlds. It was also appreciated that Jupiter and Saturn do not have large tilts because they are just too massive to be rocked by anything much smaller. But what about Uranus and Neptune? The spin axis of Uranus is tilted 98 degrees from an "upright" position, and Neptune is tilted over by about 30 degrees. It is not easy to tilt spinning tops like these – each has a mass about 15 times that of Earth (Figure 6.1). The only really viable way these tilts could have been generated would be if Uranus and Neptune had collided with some very large competitors to their respective thrones during accretion. Large is almost an understatement here. To tilt Uranus and Neptune, calculations show that the ancient impactors must have been Earth-sized or larger.

Pluto and Charon, S. Alan Stern and Jacqueline Mitton
Copyright © 2005 WILEY-VCH Verlag GmbH & Co. KGaA, Weinheim
ISBN: 3-527-40556-9

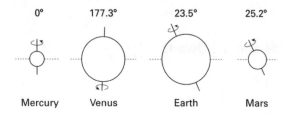

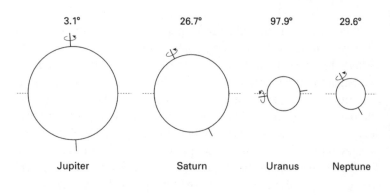

122°

Pluto

Fig. 6.1: The degree to which the rotation axis of each planet is tilted with respect to the plane of its orbit around the Sun. A tilt greater than 90 degrees means that the planet's rotation is *retrograde* – that is, in the opposite sense to its motion around the Sun. The planets in our solar system with retrograde rotations are Venus, Uranus, and Pluto. (The giant planets are shown scaled down by a factor of five relative to the terrestrial planets.)

One can use the particle in a box (PIB) method described in Chapter 5 to estimate how many large impactors would have been needed in the Uranus–Neptune zone to ensure a high probability that Uranus and Neptune would collide with objects of an Earth mass or larger. Such a calculation reveals that there must have been between at least several, and perhaps as many as ten roughly Earth-mass or larger objects in the region where Uranus and Neptune formed in order to have a significant chance of tilting Uranus and Neptune. This is very strong evidence that the outer solar system was significantly more crowded in its early days than it is now.

Yet another line of evidence indicating the presence of a higher population of ancient planetary bodies in the outer solar system comes from studies of Neptune's largest moon, Triton (Figure 6.2). Triton is similar to Pluto in several striking respects: its diameter of 2700 kilometers (1690 miles) is barely different from Pluto's (2300 kilometers or about 1450 miles); further, Triton and Pluto each have densities of almost precisely twice that of water, and each displays a surface covered in nitrogen, carbon monoxide, methane, and water ice, with nitrogen dominating the mix. Indeed, Pluto and Triton are more similar to one another than are any two planets in the solar system.

Fig. 6.2: A view of Neptune's moon Triton assembled from 14 separate images obtained by *Voyager 2*. This mosaic shows a variety of different kinds of terrain. The streaked south polar cap is apparent at the bottom. The dimpled surface visible towards the upper right has been likened to the skin of a cantaloupe melon. To the right there is a region of smooth plains and low hills which is the most densely cratered part of Triton seen by *Voyager 2*. (NASA)

Yet the most amazing thing about Triton is not its similarity to Pluto but the fact that its orbit around Neptune is backwards relative to

Neptune's rotation and to the orbits of all of Neptune's other satellites. This so-called "retrograde" orbit is ironclad evidence that Triton was gravitationally captured by Neptune, and therefore that Triton was once a planet in orbit about the Sun, just as Pluto is now.

Triton's capture from solar orbit has been studied by some of the best planetary theorists in the world, including Peter Goldreich and Bill McKinnon. The trick needed in finding a way to get Neptune to capture Triton is to bring Triton close to Neptune and, at the same time, dissipate sufficient orbital energy and momentum so that Triton enters a stable orbit around Neptune.

One way to effect Triton's capture is a perfectly aimed, grazing collision by Triton through Neptune's atmosphere. (Not too shallow or Triton would escape. Not too deep or Triton would never re-emerge into orbit.) Alternatively, while passing through the Neptunian system, Triton might have just happened to have struck a satellite of Neptune, shattering the luckless moon, but in the process reducing its own energy and momentum enough to achieve orbit about Neptune. A third alternative envisages that the capture was aided by gas drag from the tenuous nebula around Neptune during its formation.

What is clear from these ideas[32] is that, regardless of whether Triton entered Neptune's atmosphere "just right," just happened to hit a satellite of Neptune, or passed by during the brief period when Neptune's atmosphere was grossly extended, something improbable occurred. And just as was the case with the Charon-progenitor finding Pluto to make the binary, the best way seen to avoid the long odds of these improbable events is to postulate that many Triton-like bodies must have once been extant in this region of the solar system.

Putting together the evidence from the Pluto–Charon binary, the tilts of Uranus and Neptune, and the capture of Triton, it seems that the outer solar system is telling us something. At Uranus, at Neptune, and at Pluto as well, there are the smoking guns of long-past events that indicate that the ancient solar system made small and moderate sized planets in great numbers. Were this not the case, the combined odds of our seeing a tilted Neptune, *and* Triton in a retrograde orbit, *and* a Pluto–Charon binary would be too long to take seriously.

The conclusion one thus reaches is that there must have been hundreds to thousands of Pluto-sized and larger bodies produced in the ancient outer solar system.

Where Have All the Pluto's Gone?

For over 30 years now, computer simulations have been used to study the final stages of outer planet formation. The most notable of the early formation simulations were performed by Victor Safronov and his colleagues in the Shternberg Institute in Moscow, and by the team of Julio Fernandez, a university professor in Montevideo, Uruguay, and Wing Ip, who was then in at the Max Planck Institute for Aeronomy[33] in Lindau, Germany.

These groups demonstrated some important things about the formation of the giant planets. First, they showed that the giant planets are sufficiently far from the Sun, and so massive, that they gravitationally eject far more planetesimals from their formation zones than they actually accrete. A key reason for this is simply that objects on crossing paths are more likely to pass close to, but miss, a planet than to actually hit it. Objects on such near-miss trajectories receive a strong orbital deflection[34] oftentimes causing them to be ejected from this distant region of the solar system. With literally trillions of planetesimals orbiting between Jupiter and Neptune, the mêlée around each budding giant planet must have been tremendous, with every manner of rock, boulder, collisional shard, and planetesimal that passed sufficiently close having good odds of being ejected from the region altogether.

Once the giant planets begin reaching their present-day masses, there was virtually no refuge in the outer solar system from their gravitational reach. In fact, studies of the orbital dynamics of the outer planet region by workers like Hal Levison and Martin Duncan show that within just 10–100 million years after reaching their present-day masses, the giant planets emptied virtually the whole region between 4 and almost 34 AU from the Sun of everything but one another.[35] Most of what was not emptied on this timescale was cleared out in the first billion years after the giant planets formed.

Interestingly, Jupiter and Saturn are so massive that almost all of the objects that they scattered outward were ejected completely from the solar system. However, because Neptune and Uranus have far smaller masses than Jupiter and Saturn, they ejected far fewer bodies, and instead deposited between 10% and 30% of the planetesimals and debris that they scattered into distant, weakly bound orbits beyond Neptune.

Numerical simulations show that most of the planetesimals and smaller debris launched to these distant orbits by Uranus and Neptune now populate a zone that stretches a thousand times farther away than the giant planets. This region is called the Oort Cloud, after the Dutch astronomer Jan Oort, who first recognized back in 1950 the evidence for it in the pattern of long-period comet orbits. This gigantic cloud of debris from the era of giant planet formation may contain as many as a 10^{12} (a million million) ejectees a kilometer or more in size. Some, it seems, may be as large as Earth!

Of course, this all begs a question about Pluto and Triton. Given that Uranus and Neptune were so efficient at gravitationally sweeping up or sweeping out everything between and around their orbits, why do Pluto and Triton remain? They survive only because they are resident in stable *dynamical niches*, where they were protected from ejection. These niches are the astronomical equivalent of the La Brea tar pits in California. Whereas the tar pits preferentially preserve old saber tooth tiger bones that would otherwise have been long lost to erosion, the dynamically stable niche regions of the outer solar system preserved long-extinct relics from the formation era of the giant planets. Triton's niche is its safe orbit around Neptune. Pluto's niche is its cozy 3:2 resonance with Neptune.

The Edge

When the story outlined above, arguing for a thousand or more ancient ice dwarf planets in the ancient outer solar system, hit the streets as a scientific paper by one of us (Stern) in the spring of 1991, reaction was, well, mixed (Figure 6.3). Some thought it just the right answer to the long-standing lack of context for Pluto (and Triton). Others thought it a speculative "fairytale," supported only by circumstantial evidence. Further, it was pointed out, if the theory was right, then the direct evidence to confirm it had been neatly removed, billions of years ago. "How convenient," someone said.

Even for scientists, seeing is the key to believing, so the question became how to find concrete evidence of the multitudes of ice dwarfs that were thought to have formed along with Pluto. Looking to the far away Oort Cloud did not seem promising. Calculations showed that a Pluto-like object ejected into the Oort Cloud would be a billion

Fig. 6.3: The cover of the weekly American science magazine *Science News* on September 21, 1991 proclaimed the new idea that Pluto may once have been just one in a whole population of icy dwarf planets that formed in the early outer solar system.

times fainter than Pluto – so faint that even the very largest telescopes could not hope to find it. Worse still, such an object would only move 1 arcsecond every 4 years, making it almost indistinguishable from stars.

There had to be an easier way, and it came when it was suggested that some of these objects might be found in another stable region that might lie as close as a billion miles beyond Neptune.

The region beyond Neptune, which astronomers call the trans-Neptunian zone, has been a frontier begging for exploration since Kenneth Edgeworth (in 1943 and 1949) and Gerard Kuiper (in 1951) wrote seminal papers pointing out the same great puzzle of the region. What intrigued Edgeworth and Kuiper was the sharp border that the solar system presents at 30 AU, where Neptune orbits. Inside 30 AU there are the planets, but beyond 30 AU, there appeared to be, well, nothing but tiny Pluto.

And why should that be? Why, Kuiper and Edgeworth asked, should the planetary system end so abruptly? Why did it not instead peter out more gradually with distance? Why, they each asked, was there not at least a debris field of planetesimals left over from outer planet formation? Why was there only Pluto beyond Neptune?

Edgeworth and Kuiper's ruminations were on the trail of something big, but few mid-century astronomers appreciated that. In part this was because the technology really did not allow the question to be easily addressed. In part it was also just that the *lack* of material beyond Neptune just did not strike many people in 1950 as very odd. Edgeworth and Kuiper felt that the almost perfect emptiness beyond the last planet was bothersome, but they could not drum up much enthusiasm or interest about it among the larger astronomical community. They were too far ahead of their own time in this pursuit.

The beginnings of paradigm shifts are often like that. In fact, Edgeworth and Kuiper's early musings are reminiscent of the rumblings in late nineteenth century physics. Sometime around 1890 it was commonly said that physics was almost completely solved and that once the photoelectric effect and the Michelson–Morley speed of light experiments were explained, physics, as a field, would be all but complete. Little did the purveyors of this notion know that relativity, quantum mechanics, particle physics, fusion, and a host of other discoveries lay just over their horizon. So too, the astronomical community of the 1940s and 1950s did not realize that the trans-Neptunian region was a whole new frontier for planetary science – if one looked hard enough to see what was there.

The Hunters

As we just noted, Kuiper (Figure 6.4) and Edgeworth wrote down their ideas about the puzzle they felt the empty trans-Neptunian region presented, but the technology of the times prevented much from

Fig. 6.4: Gerard P. Kuiper (1905–1973), after whom the Kuiper Belt is named, in about 1955. He was an early Pluto pioneer and one of the most distinguished planetary scientists of his time. (Yerkes Observatory)

coming of their speculations. So like the two men themselves, their ideas soon passed.

Time also passed – over four decades of it, in fact. There came and went a cold war, and several hot ones. There came and went Apollo, the Mariners, the Vikings, and the Voyagers. There came and went Elvis, James Dean, the Beatles, and Curt Cobain. There came commercial jet transport, weather and communications satellites, and a computer revolution that continues without apparent bound. Throughout, the question Edgeworth and Kuiper posed remained completely unresolved, and still largely unappreciated – until one late August day in 1992.

The place, as it often is in modern astronomy, was atop Mauna Kea, nearly three miles above the shores of the Pacific, on the big island of

Fig. 6.5: David Jewitt and Jane Luu, photographed in about 1996, who discovered 1992QB$_1$, the first Kuiper Belt Object (KBO) to be identified. (J. Mitton)

Hawaii. The players were Dave Jewitt, an astronomer at the University of Hawaii, and his former graduate student Jane Luu, who was at the time a postdoctoral researcher at the University of California at Berkeley (Figure 6.5). Jewitt and Luu wanted to answer Kuiper and Edgeworth's question: is the solar system beyond Neptune empty save Pluto, or not?

Their tool was a 2.2-meter telescope, almost exactly the size of the Hubble Space Telescope, but located on the ground. Their search began in the late 1980s with the advent of a new generation of sensitive charge-coupled device (CCD) cameras that could obtain images of the sky with a clarity and depth few had imagined possible even in the early 1980s.

Jewitt, a British expatriate and professor at the University of Hawaii, was 35. Luu, a former Vietnamese refugee, who as a girl literally escaped in the last wave before the fall of Saigon in the Vietnam War, was still in her twenties. In 1982, when he was still a Caltech graduate student, Jewitt had used the telescopes atop Mauna Kea and a then state-of-the-art CCD camera to spot before anyone else the faint nucleus of comet Halley returning toward the Sun. In the intervening decade, Jewitt, with students like Luu, had been pushing CCD cameras further and further in the study of faint objects in the deep outer solar system. During that time CCDs had grown larger, making it possible to capture a broader search swath with every image. CCDs had also grown more sensitive, largely as a result of engineering advances that reduced their electronic noise.

Jewitt and Luu were not alone in undertaking such a search. Accomplished astronomers like Hal Levison and Martin Duncan, Anita Cochran, and Tony Tyson were trying the same thing. All of them knew it would not be easy.

Clyde Tombaugh's search for Pluto had long before proven that nothing as bright as Pluto (15th magnitude) was out there, moving slowly against the stars. And in the 1970s, US Naval Observatory astronomer Charlie Kowal had conducted another, deeper search, this time penetrating to almost 19th magnitude – 100 times fainter than Pluto. In fact, despite years of effort and a search that wrapped its way halfway around the ecliptic, Kowal found only one object, Chiron, orbiting primarily between Saturn and Uranus (Figure 6.6). At the end of his search, Kowal estimated there might at most be a

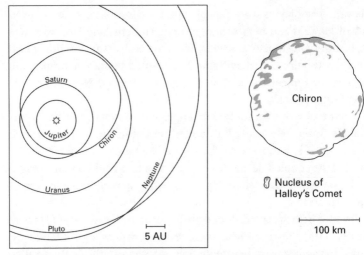

Fig. 6.6: Chiron, now considered the prototype of a class of solar system bodies known as Centaurs, seems to be intermediate between a comet and an ice dwarf planet like Pluto. Chiron's orbit relative to the outer planets is shown on the left. On the right it can be seen that Chiron is much larger than a typical cometary nucleus. The nucleus of Halley's Comet is shown for comparison. Nevertheless, Chiron develops a diffuse coma around itself – a characteristic associated with comets rather than planets or asteroids.

handful more objects 100 or so times fainter than Pluto, but anything else out there – if there *were* anything else – was fainter still.

With that knowledge in hand, the trans-Neptunian search teams began their deep surveys at the end of the 1980s. All four groups used the strategy of searching the opposition point pioneered by Tombaugh six long decades before. Initially, each group reported negative results, meaning they had found nothing so far, and would have to search more sky, and to fainter limits, if a discovery was to be made. Some groups dropped out. Others persisted in the search. None worked harder than Jewitt and Luu.

Enter Smiley

In late August of 1992, after four years of work, Jewitt and Luu were rewarded with the detection of a faint ember, moving slowly against the myriad stars on their CCD images. The object they had spotted was thousands of times dimmer than Pluto, and more than 50 times fainter than anything Kowal could have detected in his photographic search 15 years before.

Like Tombaugh over 60 years earlier, Jewitt and Luu wanted to check their work before announcing it publicly. Sure-footedness, they knew, is far superior to the embarrassment of a public retraction. So, over the next few days, before the Moon came back to wash out the night sky and halt their work, the pair gathered additional observations of the particular patch of sky surrounding their find. The repeated imaging they did showed the object was not only real but, importantly, crawling across the ecliptic star fields at just the rate expected for a trans-Neptunian object.

One of the first people let in on Jewitt and Luu's find was astronomer Brian Marsden of the Harvard-Smithsonian Observatory in Cambridge, Massachusetts. Marsden is one of the world's most careful and meticulous orbit plotters and, in his role at that time as the Director of the International Astronomical Union's (IAU) Minor Planet Center, he was the man to judge when the distant object's reality was sufficiently well established to become public knowledge.

Jewitt and Luu determined the distance to their prey from its rate of motion on the sky. The faint, slowly moving object they had found was most likely about 43 AU from the Sun, fully 2 billion kilometers (1.2 billion miles) beyond Neptune. Jewitt and Luu found that, if they assumed the object's reflectivity was low, as expected for an ancient, radiation-battered surface, then its diameter must be in the neighborhood of 260 kilometers (140 miles). That is more than 10 000 times the size of comet Halley – and about ten times smaller than Pluto. At this size, their nomad object was probably big enough to become a kind of lumpy sphere, indicating it was near the borderline at which objects reached the clear hallmark of planethood: roundness. No doubt about it, the distant wanderer could be called a minor, or better yet, a dwarf planet.

Marsden evaluated the data Jewitt and Luu had collected and on September 14, 1992 issued a bulletin, called an *IAU Circular*, to astronomers around the world. Such IAU bulletins are issued several times a month to inform rapidly the worldwide astronomical observing community of discoveries that would benefit from rapid follow-up observations (e.g., supernovae or new comets).

Marsden was careful to couch his remarks about the trans-Neptunian discovery in appropriate language that would alert other astronomers of the find, without overstating the case that the first-ever Kuiper Belt Object had actually been confirmed. Marsden's announcement read, in part:

> "D. Jewitt, University of Hawaii; and J. Luu, University of California at Berkeley, report the discovery of a very faint object with very slow (3 arcseconds/hour) retrograde near-opposition motion, detected in CCD images obtained with the University of Hawaii's 2.2-m telescope at Mauna Kea. The object appears stellar in 0.8 arcsecond seeing, with an apparent magnitude $R = 22.8 \pm 0.2 \dots$ Computations $\dots$ indicate that $1992QB_1$ is currently between 37 and 59 AU from the Earth but that the orbit (except for the nodal longitude) is completely indeterminate. Some solutions are compatible with membership in the supposed Kuiper Belt, but the object could also be a comet in a near-parabolic orbit."

As Marsden's announcement indicated, there was one more hurdle to pass before Jewitt and Luu could be sure that their find was indeed a true inhabitant of the trans-Neptunian region. They had to eliminate the possibility that the object was on an elliptical orbit plunging toward the Sun through the trans-Neptunian region, as opposed to permanently circling there. This would require repeated monitoring of $1992QB_1$'s movements over a period of three to six months (Figure 6.7).

According to the strict nomenclature rules of the IAU, Jewitt and Luu's find, known as $1992QB_1$, could not be named until its orbit was precisely established. $1992QB_1$ not a very romantic name,[36] but IAU nomenclature rules for objects in the solar system clearly state that nominations for a personal pedigree – a name – on any newly discovered object will not be accepted until the object's orbit is accurately known. The reason for this is to prevent named objects being lost owing to poor knowledge of their orbits. Jewitt and Luu knew they would have to wait for an orbit good enough for the IAU to be available, but they told anyone who asked that they had already selected the name they planned to give their find. It would be *Smiley*, after their favorite spy from John Le Carré's popular novels.

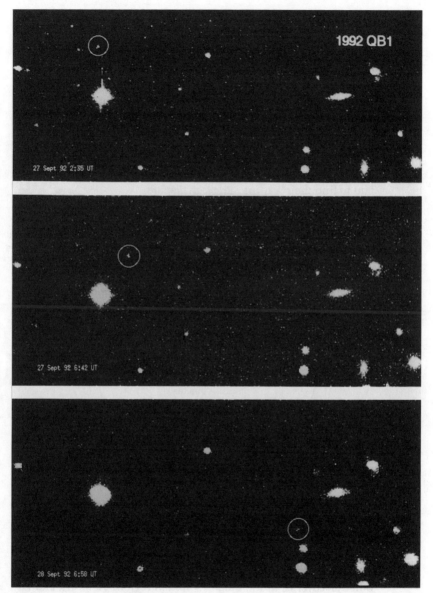

Fig. 6.7: This series of three images of the Kuiper Belt Object 1992QB$_1$ was made on September 27 and 28, 1992, about one month after its discovery. The images were taken with the European Southern Observatory's 3.5-meter New Technology Telescope in Chile. The faint (23rd magnitude) image of 1992QB$_1$ is circled on each frame. It was moving by about 75 arcseconds per day against the stars. (European Southern Observatory)

Pluto's Kin

Despite the fact that "QB$_1$'s" orbit had not been confirmed to be in the Kuiper Belt, the news of the discovery moved rapidly from astronomers to the astronomical press to the wider news media. Within a day of Marsden's IAU Circular, QB$_1$ was on the front pages of the *New York Times*, *The Times* in London, and dozens of other newspapers around the world. Before the month was out, the story was on CNN and in *Time* magazine. The firestorm of coverage was so intense that for a while some astronomers actually expected Jewitt and Luu to find their way into *People* magazine.

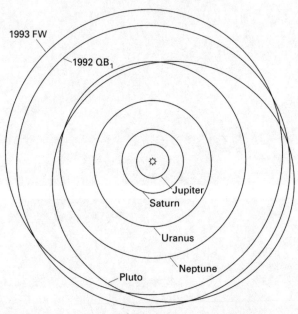

Fig. 6.8: The orbits of the first two trans-Neptunian objects (other than Pluto) to be identified, 1992QB$_1$ and 1993FW.

Finding QB$_1$ was an icebreaker and it opened up a floodgate of other discoveries in the trans-Neptunian region. As Jewitt and Luu worked to get additional observations and improve knowledge about QB$_1$'s orbit, they continued their search for other trans-Neptunians. Less than six months after QB$_1$ was announced, Jewitt and Luu found a second trans-Neptunian, designated 1993FW (Figure 6.8). Then in the late summer of 1993 they found two more: 1993RO and 1993RP.

The same week, a team from the UK led by Iwan Williams announced they had also begun detecting trans-Neptunians, with the discoveries of 1993SB and 1993SC. By late 1994 the count had reached 15, and by the end of 1996 no less than 40 of Smiley's brethren had been nabbed. The Kuiper Belt really existed. By early 2005 the total number of Kuiper Belt Objects (KBOs) discovered had topped 1000! Some looked to be as big as 1500 kilometers (over 900 miles) in diameter, rivaling Pluto's size.

Astronomers studying these objects found they came in a variety of colors, from gray to deep red. Likewise, they came in sizes that ranged down to as little as 10 kilometers, and with rotation periods ranging from a few hours to many days. Most amazingly, beginning in 2001, KBOs began to be discovered with large moons, oftentimes mimicking the Pluto–Charon binary in miniature.

But perhaps most stunning of all was that the 1000+ KBO discoveries made by early 2005 were just the tip of an iceberg. Consider why. All of the hundreds of images taken in the search had covered only a few square degrees of sky, which is just a minuscule fraction of the thousands of square degrees near the ecliptic where such objects are to be found. Comparing the area that had been searched to the area of the entire ecliptic region, it was clear that about 100 000 KBOs of 100 kilometers diameter or larger must be orbiting in the region between 30 and 50 AU from the Sun. Pluto is not alone! In fact, it is orbiting among a throng of smaller kin left over from the era of outer planet formation (Figure 6.9).

Meet the Plutinos

Even when there were only 100 KBOs known, it was already obvious that their orbits fell into several distinct subgroups (Figure 6.10a, b and c). Most KBOs are in only moderately elliptical orbits that are entirely contained in the region between 30 and 50 AU from the Sun. This group is often called the Classical KBOs (CKBOs). Others, however, follow extremely elliptical orbits that range between two and over 10 times farther out – some reaching over 1000 AU from the Sun at their most distant points. This group is called the Scattered KBOs (SKBOs) because it appears they were ejected to these distant orbits by Neptune's gravity.

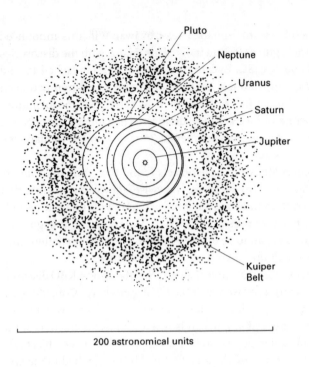

Fig. 6.9: The Kuiper Belt (or Disk) is thought to occupy a region extending approximately from the orbit of Neptune to a distance of 100 astronomical units (AU) from the Sun. It is shown here schematically face on and from the side.

Most interestingly from our Plutocentric perspective, however, is a distinct subgroup of objects that are in the same 3:2 mean motion resonance with Neptune that Pluto itself is in. Because these objects have orbits so much like Pluto's, they have been dubbed the "Plutinos." It is estimated that about 10% of all KBOs are Plutinos. Pluto is the largest object in the group.

How did the 3:2 resonance come to be populated with so many bodies? Renu Malhotra of the University of Arizona, building on

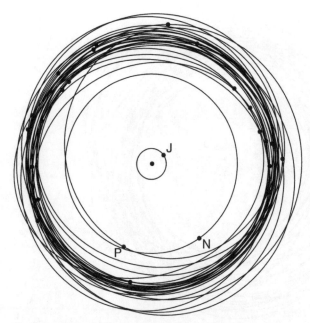

Fig. 6.10: Orbits of different classes of Kuiper Belt Objects. (a)
The orbits of some typical Classical Kuiper Belt Objects. The small
circle marked J is the orbit of Jupiter. The orbits of Neptune and
Pluto are marked N and P.

previous work by Julio Fernandez, has proposed and demonstrated
the viability of the most likely explanation. As a result of angular
momentum exchange with planetesimals during the accretion of the
planets, the outer planets underwent a migration.

As the giant planets ejected the comets to the Oort Cloud, the sizes
of their orbits changed. In particular, numerical simulations reveal
that Jupiter probably moved inward about one half of an astronomical
unit, but Saturn, Uranus, and Neptune moved outward. Neptune
may have moved as much as 5–10 AU.

As Neptune moved outwards – slowly mind you, for this process
took many millions of years – its mean motion resonances were
pushed outwards with it. Malhotra showed that the resonances sweep
up many of the objects they pass. The 3:2 resonance is a particularly
strong one, so it caught large numbers of small bodies, including
Pluto and thousands of Plutinos as well.

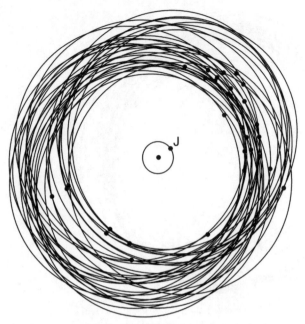

Fig. 6.10: Continued. (b) The orbits of some typical Plutinos, including Pluto itself, which is indistinguishable from the others. The orbit of Jupiter is labeled J.

Interestingly, it turns out that the migration of Pluto and the Plutinos amplifies their orbital eccentricities and inclinations, thereby explaining why these orbits are today so egg-shaped and tilted relative to those of the major planets. In fact, the observed orbital eccentricities and inclinations can be used to estimate how far these bodies were dragged with Neptune as it migrated. An important implication of the migration theory is that, while Pluto and the Plutinos formed farther out than the giant planets, this occurred somewhat closer to the Sun than they are now, most probably in the region between 20 and 30 AU from the Sun.

The Meaning of Missing Mass

One interesting clue to the origin of the Kuiper Belt, and Pluto, is that there does not seem to be enough mass in the region to build objects anywhere near the size and mass of Smiley (not to mention Pluto, a thousand times larger still) in the age of the solar system.

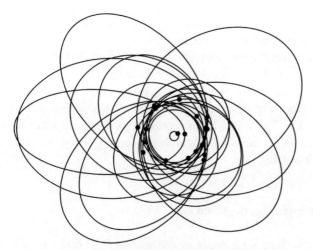

Fig. 6.10: Continued (c) The orbits of typical Scattered Disk Objects. The small circle at the center is the orbit of Jupiter. Note that the scale of this diagram is different from those of (a) and (b). The distance encompassed by this diagram is about 400 AU. (From J. Davies, *Beyond Pluto*, based on data from Chad Trujillo. Courtesy John Davies and Cambridge University Press)

This finding clearly implies that the KBOs and comets seen there today are just the remnants of a formerly much larger population of objects that was present in the distant past, before Neptune grew to its present size and the clearing of the trans-Neptunian region.

In fact, the most recent work on the clearing of the trans-Neptunian region shows important parallels with the clearing of the asteroid belt by Jupiter. In both cases, the clearing process involved both orbital perturbations and a dramatic increase in collision velocities. This converted a formerly growth-nurturing environment to a Darwinian shooting gallery, pervaded by high-speed collisions that stopped growth dead in its tracks. As a result, both the asteroid belt and the Kuiper Belt appear to be regions where a larger planet might have grown, had it only won a race against the giant neighbor we see today.

The best estimates of the combined mass of all the objects in the trans-Neptunian zone today (including comets, KBOs, intermediate-sized bodies, and Pluto), is about one-tenth of an Earth mass. By contrast, simulation results suggest that some 15 to perhaps 50 Earth masses of material were needed in the region in order for objects hundreds of kilometers in size, like the KBOs, to have accreted in less than the age of the solar system.

If that was in fact the way the trans-Neptunian region of the outer solar system was populated, then the abrupt "edge" that Kuiper and Edgeworth first noticed at Neptune was not originally an edge at all, but simply something that evolved over time as Neptune depleted this region. The KBOs and Pluto represent the remnant clues to the ancient era when the whole trans-Neptunian region was far less empty than it is today.

Hindsight 20/20: Lo, the Misfit Fits

With the advantage of hindsight we can now see that Pluto's discovery in 1930 presaged the discovery of the vast debris field of the Kuiper Belt that occurred in the 1990s. Because Pluto was (and is) so much larger, and therefore so much brighter, than any of its present-day kin, it *appeared* to be alone in the yawning vacuum beyond Neptune.

For 60 years Pluto's small size, unusual orbit, and apparent isolation in the region beyond Neptune fooled us into trying to explain Pluto as an oddity, when in fact it was a beacon, showing the way toward a then unseen flock of smaller objects quietly plying the backwoods of the planetary system.

When in the 1970s Pluto's clockwork resonance with Neptune and its double-planet nature became known, it precipitated a crisis in understanding that eliminated all of the Lyttleton-like Pluto-as-a-freak theories, in favor of much less *ad hoc*, and far more exciting possibilities. That shift in thinking began in earnest when Charon's true size was realized, forcing new thinking about how a double planet could be made, and so sparking the idea of a giant impact for Pluto. The giant impact, in turn, led to the idea that the ancient outer solar system was littered with lots of Plutos and mini-Plutos, a population often called the ice dwarfs.

Then, when the technology of CCDs made possible deep enough searches to discover Pluto's kin, Pluto's true relationship to the outer solar system was finally seen clearly. The discovery of the Kuiper Belt washed away forever the old problem of explaining Pluto as a misfit.

In point of fact, we now see that Pluto was probably built in a region containing enough building blocks to have made at least one more large planet, which never formed, and that the key to Pluto's small size is most likely the influence of "nearby" Neptune, whose own growth induced a gravitational stirring of the nest where Pluto and its cohort embryos were growing. That stirring resulted in both the collisional battering that whittled away the massive reservoir of material that once populated the 30–50 AU zone, and the rendering stillborn of the dwarfs that were growing toward full planethood in this region of the solar system. Pluto and the smaller remnants of that population that we observe today serve as testimony to the vagaries of planet formation, and as a laboratory to probe what planet formation looks like when arrested in mid-stride.

Although we are sure now that Pluto formed amidst a vast population of smaller bodies, the main thing we do not know is how common Pluto- and Triton-sized objects were in those early days. It seems that something like at least a hundred, and perhaps a few thousand, objects of Pluto's size and larger were needed if Uranus and Neptune were to be tilted over, Triton captured, and the Pluto–Charon binary created. Will we someday find such bodies farther out, in the distant Oort Cloud perhaps? Stay tuned. The situation is unfolding before our very eyes.

7
Everest

"They say you want a revolution, well, you know, we all want to change the world." – The Beatles

With this chapter and the next we turn a corner, moving away from a description of what things have been learned about the Pluto–Charon binary to the planning for a spacecraft mission to reconnoiter this fascinating pair of frozen worlds. But before we move on to the subject of how a mission to Pluto came to be, it is useful to remember what past spacecraft missions to planets have taught us, and to summarize what we know – and what we yearn to know – about Pluto and Charon.

Cradle Vista

Perhaps the main lesson of planetary science is just how limited the field was before space exploration became reality. The limitations of that purely astronomical era were not the size of the telescopes, but a series of other factors. First, there was the comparatively primitive nature of the instruments and detectors available to put at the focus of ground-based telescopes. By some accounts, the advances in instrument and detector technology make each telescope 500 000 times more effective today than in the 1950s and 1960s. Second, the computers to analyze the data and model the results were scarce and, even when they were available, they were primitive. Third, astronomers were blind to some of the most valuable parts of the electromagnetic spectrum, including the rich terrain of the ultraviolet and infrared regions. And fourth, there was the maddening turbulence of the Earth's atmosphere, which reduced telescopic resolution so far below the theoretical limits of the telescopes that the planets appeared only as fuzzy orbs and their satellites as simple pinpoints. Without the ability to *see* what the surfaces of the planets looked like at mod-

Pluto and Charon, S. Alan Stern and Jacqueline Mitton
Copyright © 2005 WILEY-VCH Verlag GmbH & Co. KGaA, Weinheim
ISBN: 3-527-40556-9

erate resolution, planetary astronomers were less well off than the proverbial blind man exploring an elephant by touch alone.

Given these severe limitations of mid-twentieth-century astronomical tools and techniques, it is not much of a mystery why the solar system we know today seems so very different from the one that Tombaugh, Kuiper, and colleagues knew before spaceflight, in the 1950s.

Yes, their solar system consisted of nine planets, but more than half were so poorly understood that even the length of their days were uncertain. There was not a single definitive fact about the composition of any planetary satellite, save our Moon and Saturn's Titan. So too, no one knew whether any, all, or some of the planets (except Earth) had magnetic fields. And the rings of Jupiter, Uranus, and Neptune were so unsuspected that reputable astronomers critiqued cartoons for showing rings around planets as a common occurrence.

So too, prior to the dawn of the Space Age, some responsible astronomers thought Mercury had a substantial atmosphere, and that Venus was quite plausibly habitable. Others imagined Mars might be teeming with an alien biology, and that the asteroids might be an exploded planet. Pluto was widely thought to be a rogue misfit...

The list goes on, but we shall not. The solar system known to the generation of astronomers and enthusiasts who toiled in the years between the Second World War and the dawning of the Space Age was as far removed from the solar system we know today as a horse and buggy is from a car.

The scientific revolution brought about by the robotic exploration of the planets has, most fundamentally, been a revolution in perspective. That exploration taught us that the solar system is far more diverse, more subtle, more complex, and more beautiful than anyone suspected. Exploration taught us that the richness of Nature is not limited to rainforests or summer skies. And exploration taught us that most of our cherished, arm-chair understandings of the planets and their retinue of lesser worlds were but a child's view of our home planetary system.

An Ornament Against the Deep

Despite the fact that we know remote exploration by telescope often underestimates the richness and intricacies of Nature, it is useful to summarize what has been learned so far about Pluto–Charon. Much of what we know has come about because we have the technological tools to probe an object so faint and far away but, in addition, much has come through the good luck of having both the rare mutual events and a once-every-248-year perihelion passage to observe. It is from those good fortunes that a stick-figure portrait of the Pluto system has been taking shape, as we now summarize.

To begin, Pluto is not just another planet with a satellite, but is instead a binary planet, consisting of two comparably sized but otherwise vastly different worlds, Pluto and Charon. This strange binary nature, along with its unusual orbit, and its diminutive size, characterize the Pluto–Charon system.

The Plutonian pair orbits the Sun every 248 years, coming as close as 29.5 AU and receding as far as 49.6 AU. Their orbit around the Sun is locked in a clockwork resonance with Neptune that shows Neptune played some important role in Pluto's past orbital evolution.

The sizes of Pluto and Charon are known to an accuracy of a few tens of kilometers. We also know that the pair lies tipped over, on their side, spinning on their axes every 6.3872 days. Charon orbits Pluto in Pluto's equatorial plane, with this same period; as a result, Charon hovers over one spot on Pluto's equator. Charon never sees the other side of Pluto, and so too that "far-side" is ignorant of Charon's visage.

All searches for other satellites in the Pluto–Charon system have failed to find anything. These searches have placed strong limits on the mass and size of any satellites that might still remain undiscovered. Nothing larger than 5% of Charon's diameter – a mere 50 kilometers or so – could be hiding there, still unknown to us. Perhaps Pluto has no other companion save Charon; perhaps one or more small moons also accompany it – we do not know.

The Pluto–Charon binary probably formed through a giant collision, about 4 billion years ago. Based on the odds of such a collision occurring, and the discoveries of so many, almost Charon-sized objects in the Kuiper Belt region where Pluto orbits, it is now widely accepted that Pluto and Charon are among the largest remnants of a once-teeming population of miniature planets called ice dwarfs that were formed during the era of giant planet building in the deep outer solar system.

Let us now look a little closer at the two bodies that make up the Pluto–Charon binary. We start with the smaller of the pair, Charon. We know that Charon shows only subtle variation in brightness as it rotates, which indicates that its surface appearance is blander than Pluto's. Charon's gray color and much darker albedo than Pluto's also contribute to Charon's blander appearance. This blander appearance also extends to surface marking on Charon – for none have ever been identified, even in the best images taken. However, Marc Buie has found that Charon's surface must have some markings, for it varies in brightness by just under 10% as it presents different faces to us during its orbit around Pluto.

Spectroscopic studies have revealed that Charon's surface is covered with water ice, ammonia ice, and ammonia hydrates; other ices, or even rocky materials, may lie on Charon's surface as well, but they have not been uniquely identified yet. There is no proof (or even much indirect evidence) of an atmosphere on Charon. Based on everything that is known about Charon, which is not much more than described here and in Table 7.1, the smaller member of the Pluto–Charon binary is expected to be rather like the 1000-kilometer-wide, gray, water-ice-covered satellites of Uranus, like Ariel.

Table 7.1: Some basic attributes of Pluto and Charon.

Parameter	Pluto	Charon
Rotation period (days)	6.3872	6.3872
Diameter (km)	2360	1200
Density (g/cm^3)	2.0	1.7–2.1
Surface reflectivity (%)	55	35
Lightcurve amplitude (%)	38	8
Known surface ices	N_2, CH_4, CO	H_2O
Atmosphere	Confirmed	Doubtful

We now turn to Pluto, which is the life of the Pluto–Charon party. Over 30 years of spectroscopic studies of Pluto's surface show that it is covered with a mixture of exotic snows, including methane (CH_4), carbon monoxide (CO), and nitrogen (N_2). Although the nitrogen ice dominates Pluto's surface composition now, the proportions of the various ices change both with Pluto's seasons and its back-and-forth swings in distance from the Sun.

Pluto has an atmosphere, which envelopes it in a puffy shroud thousands of kilometers high. At the surface, the pressure of the atmosphere is just a few, or perhaps a few tens of microbars – less than 1/10 000 the pressure atop Earth's Mount Everest. Pluto's atmosphere, like Earth's, consists mostly of N_2 gas, but unlike Earth's atmosphere, Pluto's also contains significant quantities of CO and CH_4. Importantly, the CO and CH_4, though present only in trace amounts, play an important role in generating a rise in temperature of the atmosphere above the surface. There is evidence that the atmosphere may sometimes have haze layers, and that it waxes and wanes dramatically in mass as Pluto moves around the Sun. Pluto's atmosphere must also contain some as-yet undiscovered trace constituents that result from the chemistry of sunlight shining on the N_2–CO–CH_4 mixture. A similar kind of chemistry acting on the surface ices darkens and reddens Pluto's surface. As a result, Pluto's highly reflective surface sports a tinge of ruddy pink.

Pluto's brightness varies dramatically as the planet rotates, which provides strong evidence of a highly variegated surface. (And it is no wonder, with such a complex mixture of surface ices and their chemical byproducts.) These suspicions were confirmed by the maps made using data from the mutual event season that lasted from 1985 to 1990, and Hubble Space Telescope images made in the mid-1990s.

Owing to the intricate patchwork of bright and dark ices and other materials on Pluto, its ground temperature appears to vary from place to place by huge proportions – perhaps by as much as 50% from the warmest (about 60 K) to the coldest (about 35 K) spots. Such large temperature gradients, in turn, probably drive intense winds and complex circulation patterns. No doubt these too shift with the seasons, as the ices migrate and the atmospheric bulk readjusts to the rise and fall of solar heating caused by Pluto's motion around the Sun.

Pluto's density is known well enough to constrain its internal composition. It is almost certainly a mixture with 60% to 70% rocky materials, a few percent exotic ices like those detected on the surface, and perhaps 30% to 35% water ice. We suspect that Pluto has differentiated into a layered planet, but that has not been proven.

And there our knowledge comes to a rather abrupt end.

On Being There

The picture of Pluto–Charon that we have just sketched has about as much detail as astronomers might have painted of Mars in 1965 or Jupiter's moon Io in late 1977, just before each was visited by reconnaissance spacecraft. In a single day, those first spacecraft flybys of Mars and Io revolutionized our knowledge and perception of these worlds, just as did the close-up scrutiny that spacecraft provided of Mercury, Venus, Europa, Triton, and over a dozen other worlds.

The most indelible lesson of planetary science is that Nature always outperforms our expectations and, on close inspection, often shatters Earth-based paradigms. With Pluto–Charon the lesson remains the same: one must remove the curtain that distance imposes, in order finally to embrace the reality of a world. *Nothing substitutes for being there.*

Still, Pluto's distance from the Sun makes reaching it about as difficult a technical challenge as the solar system ever presented us with. Just consider the things its great distance implies: long flight times to a dim and far away place where solar cells will not work and communications are a thousand times more difficult than at Mars or Venus. No wonder it has been said that Pluto is the Everest of solar system exploration.

The Everest analogy aside, the astronomers who spent the late 1970s and then the 1980s discovering what a fascinating and important planet Pluto is for planetary science, were not daunted. And as the 1980s began to slip away into history, a small but determined band of Pluto's best and most dedicated explorers decided to grasp the bull by the horns and try to bring about a space mission to reconnoiter Pluto and Charon. They knew that it would not be any easy task – for the great ship NASA turns slowly, at best, particularly in recent times.

A Gathering, a Start

Humankind's journey to Pluto began at a red-checkered table in small Italian restaurant in Baltimore's Little Italy, one cool evening in mid-May of 1989. A dozen planetary astronomers, fresh from a symposium summarizing the advances made in the study of Pluto during the 1980s, were gathered at the table. The focus of their mealtime conversation was simple: how to organize enough support within the planetary science community to motivate NASA to undertake a reconnaissance mission to Pluto. The founding members of this group, which named itself the "Pluto Underground," included Fran Bagenal, Rick Binzel, Marc Buie, Bob Marcialis, Bill McKinnon, Ralph McNutt, Bob Millis, Alan Stern, Ed Tedesco, Larry Trafton, Larry Wasserman, and Roger Yelle.

The central problem, the twelve agreed, was that the reconnaissance missions to the deep outer solar system – Voyager and Pioneer – had left Pluto off their travel itinerary.

The final decision to leave Pluto unexplored by these spacecraft had come in 1979, when the Voyager project elected to skip Pluto in favor of a more in-depth exploration by Voyager 1 of the Saturn system that sent it careening out of the planetary system altogether, and a daring plan to send Voyager 2 on to Uranus and Neptune instead of Pluto (which was off in a different direction).

Back in 1979, little was known about Pluto, and the discovery of the Kuiper Belt was over a dozen years in the future. So Pluto's scientific value was not appreciated in the way it is today.

In contrast, the alternative, a close approach to Saturn's planet-sized moon Titan and the Uranus–Neptune trip offered Voyager a sure bonanza of returns and a higher probability of success (since the spacecraft would not have to survive the extra five years to reach Pluto). Right or wrong, the die was cast. Neither Voyager would visit Pluto.

By 1989, however, so much more was known about Pluto that quite a number of scientists were in hindsight beginning to regret the Voyager decisions made a decade before. Worse, by 1989, NASA's planetary program was stuck in a programmatic cul de sac, funding only giant, second-generation missions like the multi-billion-dollar Galileo and Cassini outer planet orbiters, and the billion-dollar Mars Observer. Expensive, deep-exploration missions were in style. Reconnaissance missions were out of style.

How would the young Turks of the Pluto Underground manage to redirect the inertia of NASA's planetary program to reconnoiter Pluto? They were not sure themselves, but they believed strongly that the exploration of the last planet would have the public appeal to propel itself forward – if NASA would only recommend it to Congress and the White House for approval. To accomplish that, they knew that the planetary science community in the USA would first have to reach a consensus that, in a political environment where new missions were rare, a Pluto mission deserved approval.

Over pasta, the members of the Underground agreed to begin by organizing a letter-writing campaign to Geoff Briggs, the then director of NASA's Division of Solar System Exploration. Their first goal was to raise the flag of Pluto exploration so high that it could not be ignored.

Manna, From Triton

The same summer that the Pluto Underground was organizing itself, something wonderful fell into their lap: a shift in scientific opinion that led to dramatically wider scientific support for Pluto exploration. And ironically, it was Voyager 2 that ignited that change.

By 1989, Voyager 2 was a battle-hardened veteran of planetary encounters that had revealed the secrets of more than a dozen worlds. Now, Voyager 2 was making its exit from the planetary system via a daring encounter with Neptune and its exotic system of satellites.

As it turned out, however, Neptune was not the star of Voyager 2's final flyby. Instead, it was Neptune's largest satellite, Triton, which stole the show. Voyager 2's close pass by Triton revealed it to be one of the most interesting worlds anywhere in the solar system (Figure 7.1).

Voyager 2 revealed, among other things, that Triton's surface and atmospheric composition were a lot like Pluto's. Furthermore, Voyager 2 discovered various kinds of evidence – including geysers caught in the act of eruption – that Triton is geologically alive. Given its small size (just 10% larger than Pluto) and its frigid, 40 K surface, this was a sexy scientific surprise.

The great interest in Pluto spawned owing to Triton, as well as the flood of letters to NASA headquarters by planetary scientists asking for a Pluto mission study, along with a personal visit by Bagenal and Stern, yielded quick results.

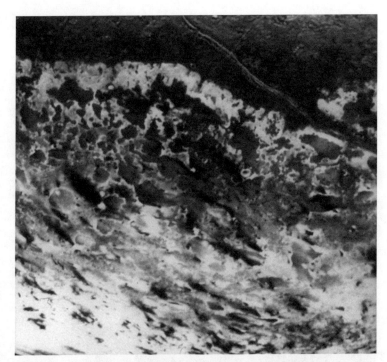

Fig. 7.1: Among the planets and moons of the solar system already imaged at close range, Triton is probably the object most similar to Pluto. This Voyager close-up of terrain near Triton's south polar region shows numerous dark plumes or streaks, which originate at very dark spots. These are thought to be vents where gas laden with dark particles has erupted into the nitrogen atmosphere from below the surface. The plumes have then been drawn out by winds. The dark streaks have been deposited on the polar terrain of bright frost mottled with darker areas. (NASA/JPL)

First Ascent

Before 1989 was out, NASA's Briggs funded a quick Pluto mission study, called Pluto 350. This effort got its name for the arcane reason that the spacecraft would have a mass target of 350 kilograms (770 pounds)... barely half the mass of Voyager.

The goal of the Pluto 350 study was to determine whether a viable scientific mission could be put together in a package this small. Given the trend of 1970s and 1980s planetary exploration missions, which featured multi-ton spacecraft, a good number of experienced insiders simply found the idea of such a small spacecraft flying across the entire solar system to be, well, ridiculous. The members of the Pluto

Fig. 7.2: Robert Farquhar, photographed in about 1996, who headed the first modern study of a mission to Pluto (Pluto 350). (Johns Hopkins University)

Underground did not much care. Perhaps it was youthful enthusiasm (few were over 35 years old then), or their political naivety, but as far as the Undergrounders were concerned planetary spacecraft were sinking under their own weight, and a small spacecraft mission to the smallest planet was just what the doctor ordered.

Briggs put NASA's Robert Farquhar (Figure 7.2) at the helm of the Pluto 350 study, and funded a design team at NASA's main planetary exploration center, the Jet Propulsion Laboratory (JPL), in Pasadena, California, to work for Farquhar. Briggs also asked Pluto Underground members Stern and Bagenal to serve as the scientific liaison to the mission, through one of NASA's planetary science advisory groups.

Farquhar, who was then a senior civil servant at NASA's Goddard Space Flight Center in Greenbelt, Maryland, was widely known for his bold and innovative mission design skills. Back in the mid-1980s, when NASA could not get Congressional approval to fund a mission to comet Halley and was faced with the embarrassment of a wholly non-US armada of spacecraft heading to the comet, Farquhar discovered a way that one of NASA's high Earth-orbiting spacecraft could be sent from Earth orbit to a comet called Giacobini-Zinner. By using a then untried slingshot trajectory past the Moon, Farquhar found that US spacecraft would arrive at comet Giacobini-Zinner 100 days before the onrushing armada was set to encounter Halley! Farquhar's plan was accepted – and it worked – making the USA the first nation to send a spacecraft past a comet.

Farquhar and the Pluto 350 design team at JPL had a considerable challenge in front of them. As we described above, for 25 years NASA's planetary mission spacecraft had been steadily growing in size. By 1990, they typically weighed in at two to four tons. Pluto 350 would have to reverse that trend.

And after only a few months of work, Farquhar and his design team managed to find a way to do just that. They packed communications, propulsion, pointing, and power systems into a compact structure weighing about 700 pounds, including the fuel needed to make trajectory corrections during the long flight to Pluto. In their design, the Pluto 350 team also left room for a powerful suite of cameras, spectrometers, and other instruments weighing about 35 kilograms (50 pounds).

Farquhar, ever the clever mission designer, realized that the cost-effective way to get Pluto 350 to Pluto was to trade a large and expensive launch vehicle capable of lofting his spacecraft directly to Pluto, for a less expensive, medium-sized launch vehicle that need only lift Pluto 350 to Jupiter. A close flyby of Jupiter could then be used to propel the vehicle onward to Pluto in just the same way that the Pioneers and Voyagers had used Jupiter to bootstrap themselves to Saturn. Very nice indeed! But it was still not nice enough for Farquhar, the Nureyev of orbital mechanics.

Early in 1990 Farquhar found that, by launching toward Venus and then making a series of Venus and Earth swingbys, Pluto 350 could be catapulted toward Jupiter after launch from one of the smallest and least expensive of NASA's existing launch vehicles – the Delta.

Farquhar's solution would cost the Pluto mission a few extra years of transit time,[37] but it would save a great deal of money that would otherwise be spent on expensive rocketry.

At the same time that Farquhar and his JPL design team were designing Pluto 350 and its clever trajectory, Bagenal, Stern, and their colleagues in the Pluto Underground were charged with defining the scientific objectives and documenting the case for such a mission. The researchers summarized the central questions that could not be answered from ground-based or Earth orbiting instruments and required on-site reconnaissance to resolve. In a report to the highest-level NASA advisory board directly responsible for planetary exploration, the Solar System Exploration Subcommittee (SSES), the Underground laid out their most burning questions:

- *How did Pluto really compare with Triton?* Were they really close cousins left over from the early era when ice dwarfs were common, or were they only superficially alike?
- *Is Pluto's surface composition as variegated as its surface markings?* That is, are Pluto's surface ices uniformly distributed, with the same amounts of nitrogen, carbon monoxide, and methane at all locations, or is the nitrogen-dominated surface spotted with a checkerboard of frozen carbon monoxide and methane ponds?
- *How deep are Pluto's ices – that is, are they simply a thin veneer lying on the surface or are they instead piled on as a deep crust?* Are Pluto and Charon's dramatic compositional differences more than skin deep? Is Pluto internally active? How much is Pluto's geology like Triton's? (After all, despite the fact the two planets have the same size and density, Earth's geology and Venus's are radically different). How does Charon's geology compare with Pluto's?
- *What is the structure of Pluto's atmosphere?* How fast is it escaping to space? Are haze layers the cause of the kinked stellar occultation lightcurve, or is the cause the atmosphere's thermal structure?
- *What are the effects of Pluto's strong seasonal changes?* Is this why Pluto's surface is so contrasty? Why are the northern and southern polar caps so different? Why is the darkest region

on the planet directly under Charon, and why is the brightest region exactly 180 degrees away from Charon?

- *How did the Pluto–Charon binary come to be?* Was it indeed created through a giant impact? Or was there something missing in our Earth-based view of Pluto, hinting that something else that occurred?

The Underground's list of questions went on, and on, and on. There was much to ask of Pluto–Charon.

Tick, Tick, Tick

While the scientists were formulating the specific questions they wanted answered by Pluto 350, Farquhar and his team realized that, because Pluto had just passed perihelion in 1989, its increasing distance from the Sun would with every passing year, make the trip time longer and the job of communicating with the spacecraft harder.

Almost simultaneously, the scientists found that time was working against them as well. Why? After perihelion, as Pluto drew away from the Sun, it would start to cool. Because Pluto's atmospheric bulk depends so sensitively on the surface temperature, the best-available computer models predicted that the atmosphere would likely undergo a rapid demise in the decade between 2010 and 2020. This demise, which experts like McDonald Observatory's Larry Trafton dubbed "atmospheric collapse," would result in a snow-out of most of the atmosphere onto the surface. If a reconnaissance spacecraft arrived after the collapse, there would not be much (or possibly any) atmosphere to study.

Moreover, they found that the onset of Pluto's deep southern hemisphere winter (driven by Pluto's 118-degree polar tilt) would be plunging more and more of Pluto (and Charon) into semi-permanent shadow, where it cannot be imaged or have its composition mapped. By 2015 almost half of the southern hemispheres of both Pluto and Charon would be shrouded in darkness – and thus hidden from view to a reconnaissance spacecraft. It would not matter if a better or faster mission arrived in 2030 or 2050: the opportunity to study Pluto's atmosphere and globally to map Pluto and Charon would be lost (Figure 7.3).

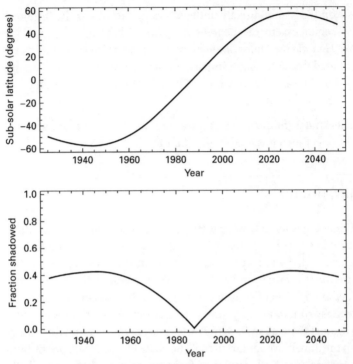

Fig. 7.3: As Pluto's aspect to the Sun gradually changes, the proportion of its surface perpetually in shadow also increases or decreases. The top graph shows the change in sub-solar latitude (where the Sun appears overhead on Pluto) over time. The lower graph shows how the proportion of Pluto's surface experiencing perpetual night changes over the same period of time. Note that almost the whole of Pluto's disk was visible in the late 1980s when the sub-solar latitude was zero. Since then, the fraction of Pluto's surface always in shadow has been steadily increasing. The longer spacecraft exploration of Pluto is delayed, the smaller the surface area visible. By 2020, only 60% of the surface will be illuminated and it will be many decades before the situation significantly improves again. (Courtesy Marc Buie)

The last time Pluto had been so favorably placed was in the middle 1700s; the next occurrence would not come until the 2230s With both the encouraging results of the Pluto 350 study and a compelling scientific case for Pluto reconnaissance in hand, it was time to act.

Farquhar, Bagenal, and Stern took the case for a small Pluto mission like Pluto 350 to the upper echelons of NASA's space science management. Simultaneously, they and the expanding roster of Pluto Undergrounders took the case up through the scientific review committee structure that NASA uses to debate and rank different mission proposals. In late February 1991, during the week of the Gulf War against Iraq, the SSES met to select its priorities for the 1990s.

A small, focused Pluto mission was popular with many members of the SSES, and its polished young chairman, Jon Lunine of the University of Arizona. But it was criticized by others, who said it had "come up through the ranks too quickly," or that it would be "too many years before the mission could return data."

But Pluto mission critics soon found themselves up against formidable odds when NASA's new chief of solar system exploration, Wesley Huntress, himself an accomplished astrochemist, announced to SSES that the scientific potential and public appeal of a Pluto mission led him to believe that such a mission should be one of the SSES's top three priorities for the 1990s.

Later in that early-1991 SSES meeting, Don Hunten, one of the most respected figures in planetary science, stood up, looked around the room, and delivered a short but effective soliloquy. The 67-year-old senior scientist almost shouted it: "I don't expect to be living when the Pluto mission arrives at Pluto," Hunten said, "And if I am living, I don't expect to be aware of the event, but this is the kind of valuable exploration we should be doing. It's important, let's get on with it." After two days of debate, the SSES wrote its report, which ranked a Pluto flyby in the highest priority category for new mission starts the 1990s. It was amazing – in just over 18 months the Pluto Underground had succeeded in raising their mission to the forefront of NASA's planned mission queue.

Less than two weeks later, NASA's Huntress formed an official science working group to guide detailed Pluto mission planning. This group was called the Outer Planets Science Working Group, or OP-SWG (Figure 7.4). The group included more than 20 of the leading American experts on Pluto and the other outer planets. Among those selected for OPSWG were Fran Bagenal, Marc Buie, Dale Cruikshank, Jim Elliot, Bill McKinnon, Dave Tholen, Larry Trafton, and Roger

Fig. 7.4: Left to right: Alan Stern (OPSWG's chairman), Wesley Huntress (director of NASA's Office of Space Science), and Richard Terrile (JPL Pluto mission scientist) at a meeting of the Outer Planets Science Working Group (OPSWG) in 1993. The OPSWG was formed by NASA in March 1991, but disbanded in mid-1996.

Yelle. OPSWG's chair, Alan Stern, and about half of OPSWG's membership were straight from the Pluto Underground. The other half of the group consisted of Voyager and Galileo mission veterans. At OPSWG's first meeting it was noted that the Pluto Underground was not very underground any longer.

In the first months after OPSWG was formed, NASA asked the group to work with JPL to study a much larger mission than Pluto 350. That mission would have used a copy of the Cassini Saturn orbiter weighing in at over 3800 kilograms (8000 pounds). Although it could have carried a whole host of sophisticated instruments and a daughter spacecraft or atmospheric probe to deepen the investigation of Pluto, the cost of this mission concept was estimated to be over $2 billion. Given the ever-tighter NASA budgets and the consequently declining resources for planetary exploration, a lead balloon could not have sunk faster.

Not surprisingly, OPSWG recommended that NASA opt for the lighter and less expensive Pluto 350 concept. For more than 20 years, NASA science working groups had designed increasingly complex

Fig. 7.5: Robert Staehle and Stacy Weinstein, photographed in about 1992, pioneers of the Pluto Fast Flyby concept. (Alan Staehle)

missions with greater and greater instrument rosters, and, by the way, increasingly greater price tags. OPSWG's decision to recommend Pluto 350 was the first time in a very long time that a scientific committee of NASA had opted for a smaller mission than NASA itself had recommended. What OPSWG did not know, however, was that a far smaller and still more radical mission concept was percolating behind the scenes at JPL.

Enter Rob Staehle and Stacy Weinstein (Figure 7.5). Staehle, a systems engineer, and Weinstein, a trajectory designer, were about as fired up a pair of young Turks as any Pluto Underground member had ever been. Back in 1991, they had discovered a new stamp set issued by the US Post Office that featured the highlights of planetary exploration. For Venus there was Mariner 2. The Mercury stamp showcased Mariner 10; a Mars stamp glorified the Viking landers.

Fig. 7.6: The stamp for Pluto in a 1991 US set devoted to the exploration of the planets by spacecraft alone declared "Not Yet Explored." It represented a challenge and a rallying point for Pluto enthusiasts determined to make a Pluto mission a reality.

The Jupiter and Saturn stamps featured the trailblazing Pioneer 10 and 11 spacecraft, and the Uranus and Neptune stamps featured the Voyager mission. The Pluto stamp said, simply: "Not yet explored" (Figure 7.6).

Staehle and Weinstein took the Pluto stamp as a call to arms. They did not know much about Pluto, or its scientific value, but they had a fire in their bellies to complete the exploration of the planets and explore this last frontier.

As it turned out, Staehle and Weinstein had never heard of the Pluto Underground – it may be a small world, but engineers and scientists run in different circles. Thinking they were alone in the world in advocating a Pluto mission, Staehle, Weinstein, and a maverick spacecraft designer named Ross Jones spent 1991 begging, borrowing, and virtually stealing JPL resources to engineer a truly tiny but capable Pluto reconnaissance mission they called Pluto Fast Flyby, or PFF (Figure 7.7).

At 140 kilograms, including both on-board fuel and its scientific instruments, PFF would weigh less than half what the already small Pluto 350 would. In fact, PFF would be so light that it could be launched directly to Pluto, thereby avoiding the elegant but time-consuming Venus, Earth, and Jupiter swingbys that Farquhar had designed. As a result, PFF could make the 5-billion-kilometer (3-

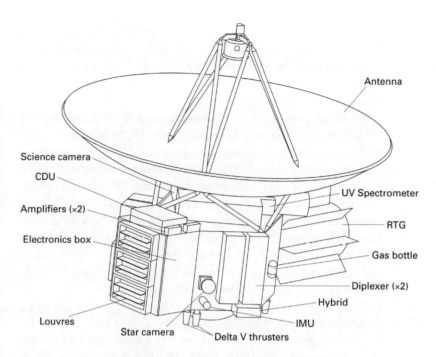

Fig. 7.7: An early version of the Pluto Fast Flyby craft. (JPL)

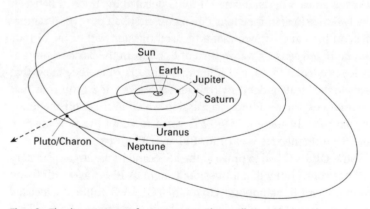

Fig. 7.8: The direct trajectory for a mission to Pluto is illustrated here. (JPL)

billion-mile) crossing in as little as 7 or 8 years, at an average speed almost twice that that Pluto 350 would achieve (Figure 7.8). By reducing spacecraft size, mass, and complexity, PFF would be significantly cheaper than Pluto 350. By reducing flight time, PFF would also reduce operations costs and allow designers to relax the stringent subsystem reliability requirements that a long, 13- to-15-year mission like Pluto 350 would impose. It was an attractive concept, but it had some important drawbacks.

The first drawback of the PFF concept was that the spacecraft could not carry a very large scientific package. After all, the package of scientific instruments on missions like *Galileo* and *Cassini* weighed more than the entire PFF spacecraft! Staehle's team initially estimated PFF might be able to carry 5 kilograms (12 pounds) of instruments. By comparison, the single lightest instrument on *Cassini* weighed almost twice that. PFF also suffered from the need to develop a great number of lightweight systems that did not then exist. Furthermore, even with lightweight spacecraft systems, there was no room for spares and many other hardware redundancies that deep space missions traditionally carried to ensure against failure.

OPSWG was directed to evaluate this concept in early 1992. OPSWG swallowed hard. The Pluto 350 call had been difficult for many of the scientists on OPSWG, but PFF was the equivalent of taking a first bungee jump. They asked, "Will this thing really work?" It was not an easy decision and OPSWG debated the tradeoff between the two missions in meetings during the first half of 1992. Pragmatists on the panel argued that a focused mission with three or four miniaturized sensors was sufficient to perform the basic reconnaissance needed at Pluto, including mapping. Further, they knew that – given the bleak budget environment at NASA – it was unlikely that the more expensive Pluto 350 mission would ever get off the drawing boards. In the end, OPSWG voted to adopt the PFF mission architecture; the vote was 3:1 for PFF over Pluto 350.

Now, OPSWG had to prove to the SSES and in the larger planetary science community that a mission as light as PFF was worth flying from a scientific standpoint. To do this, OPSWG outlined a focused set of reconnaissance objectives that it said constituted a bottom line, below which it was not worth going to Pluto. And they were not fooling around – this list of core objectives, which they dubbed Category IA, was spare indeed. It consisted of just three items, which were ranked equally. After all the scientific jargon is removed, this is what OPSWG's Category IA objectives said PFF had to accomplish if the mission was to be worth flying:

- Characterize the global geology and geomorphology of Pluto and Charon by mapping each body in its entirety at a resolution of 1 kilometer (about 3000 feet) or better.
- Obtain high spectral resolution compositional maps of the surface of both Pluto and Charon at a resolution of 10 kilometers or better.
- Accurately determine the composition and structure of Pluto's atmosphere, and characterize any haze layers in Pluto's atmosphere.

There was no fat in this list. OPSWG – Plutophiles and all – recommended that, if any *one* of these objectives could not be achieved, PFF should be abandoned. Simply put, the three legs of OPSWG's Pluto objectives formed a stable base from which to form a solid, first-order view of the Pluto–Charon system. Leaving any one of the objectives out (mapping, composition, or atmospheric assay) would result in a deeply incomplete characterization that simply was not worth the effort.

Despite this promise, tiny and spare PFF was too big a step for many and it came under heavy attack. Within the larger scientific community, many researchers said OPSWG had swayed too far under pressure by choosing a concept too bold in its engineering targets and too constricted in its scientific breadth. Others assailed PFF for its higher risk level.

Seeing that his baby was in trouble, Rob Staehle went straight to the top. Staehle learned that NASA's then chief, Daniel Goldin (Figure 7.9), was soon going to be near JPL to present a motion picture Oscar that had flown on the Space Shuttle at a small ceremony of the Motion Picture Academy of Arts and Sciences.

Staehle went to the ceremony, managed to introduce himself to Goldin, and said that his group of young engineers had a scientifically endorsed plan to get to Pluto with a revolutionary, small spacecraft, but that "The Establishment" would not allow the plan to see the light of day.

Administrator Goldin was then at the very outset of his campaign simultaneously to downsize and update NASA, and Staehle knew it. Goldin listened to Staehle, and quickly decided to make both Staehle and Pluto exploration a symbol of his new NASA. Staehle's chutzpah had seemingly paid off – Goldin went back to Washington

Fig. 7.9: Daniel Goldin, in about 1992, head of NASA between 1992 and 2001. (NASA)

and anointed the Pluto mission as his own. Needless to say, Staehle's bold maneuver peeved PFF's detractors, but the quantum leap he had made by enlisting NASA's boss directly pushed PFF into the spotlight, and toward the center of NASA's planetary mission planning for the 1990s. It seemed that a Pluto mission was now on its way toward approval for funding.

Form and Function Combined

When Dan Goldin emblazoned his sleeve with the badge of Pluto exploration, he took on a special responsibility. For the Pluto mission represented something particularly unique in all that was and is the US space program: For many, the idea of a mission to Pluto, to the very frontier of our solar system, rekindled something akin to the spirit of hope and unbridled horizons that they remembered from the explorations of Apollo.

With Dan Goldin's initial support, the drive to send a mission to Pluto moved forward rapidly. Goldin poured in resources to overcome some of the mission's most daunting challenges. The funding Goldin devoted to PFF yielded miniaturized thrusters the size of a nickel, shrunk-down communications gear, and a dozen other nitty-gritty technical improvements needed to make PFF feasible.

Perhaps the most revolutionary advances of all were those in the area of science instrumentation. Based on OPSWG's IA objectives, PFF would have to pack a sophisticated camera, plus an ultraviolet spectrometer and an imaging infrared spectrometer into a spare, 7-kilogram (16-pound) package.[38] Many in the scientific community thought this was a near impossibility and they had good reason to be skeptical. PFF would have to reduce into just 7 kilograms the capabilities of a set of *Cassini* Saturn orbiter instruments with a mass of 108 kilograms. Further, whereas the *Cassini* cameras and spectrometers consumed more than 90 watts and cost almost $100 million to produce, the Pluto package would have to consume less than 5 watts, and cost no more than $30 million – for two copies! It was the equivalent of packing fighter plane performance into a machine the size of a Cessna trainer and doing it for two bits on the dollar.

To really demonstrate that this could be done, OPSWG and Staehle's team asked NASA to fund engineering demonstrations of PFF instruments. So, in February of 1993 NASA announced its intention to select several proof-of-concept Pluto instruments for advanced development. Less than 90 days later, the space agency whittled down the proposals submitted by industrial, university, and government laboratories to a group of seven small teams, which it funded on near-breakneck, 12-month schedules to create working models of each micro-instrument.

NASA decided to fund two approaches to miniaturizing the Pluto payload. First, they would fund the development of a set of miniaturized cameras and spectrometers that could be flown as a set to cover OPSWG's IA objectives. Second, NASA would also fund a more revolutionary approach: combined instrument packages that would meld together the whole experiment suite into a single, sleek, ultralow-mass device.

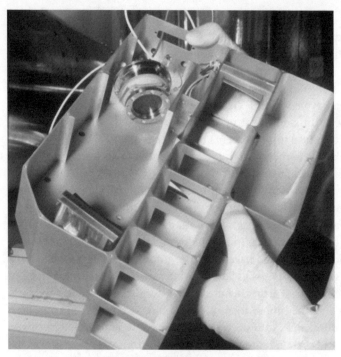

Fig. 7.10: The Highly Integrated Pluto Payload System (HIPPS) was designed to conduct a full reconnaissance of the Pluto–Charon system while being as compact as possible and keep the weight of the spacecraft to a minimum. Compare the size of the Ultraviolet Spectrograph Component (UVSC) with the human hand holding it!

The so-called *integrated* instruments combined the two spectrometers and the Pluto camera into a single unit, sharing structural and electronic resources; some even shared optics, channeling the precious light into parallel channels so that one or two telescopes would feed all three detectors. The various integrated instrument designs also experimented with exotic materials and advanced manufacturing processes, compact sensor heads, and one even featured an entire spectrometer-on-a-chip.

The most radical aspect of the integrated instruments developed for PFF, however, was what one such instrument, the Highly Integrated Pluto Payload System, or HIPPS, called a "seamless boundary between spacecraft and instrument, a payload without walls" (Figure 7.10). HIPPS was led by Alan Stern, who was by then at the Southwest Research Institute. A talented, rival team to HIPPS proposed an integrated instrument prototype called PICS, led by geologist Larry Soderblom of the US Geological Survey (USGS). PICS adopted the same philosophy as HIPPS.

The stringent mass and power targets necessary to make PFF work were tough to reach. Two of the seven selected instrument development teams failed. But five teams succeeded, by turning paper concepts into working laboratory models that met all of the performance criteria necessary to achieve OPSWG's IA science objectives. With these instruments, it appeared that PFF could expect to return more information, more detailed results, and in many ways a more well-rounded view of Pluto than Voyager's heavier and more complex suite of 11 instruments could have produced.

Late in 1994 NASA's Goldin asked for an independent review panel chaired by military spacecraft designers skilled in miniaturized space missions to assess the feasibility of PFF and its miniaturized instruments. The review panel conducted an in-depth, month-long technical review and reported to Goldin with a resounding endorsement of PFF. In less than 18 months, PFF had vaulted from little more than a daring concept to a startlingly feasible set of prototypes. Staehle and his JPL team as well as the OPSWG scientists felt ready to be turned loose to build and launch PFF. But, like Pluto 350, PFF was not to be.

Some Trouble with the Ticket

PFF would involve two spacecraft launched a few weeks apart. OPSWG had demanded this in part to mitigate the risk associated with PFF's lack of internal spacecraft systems redundancy and in part because Pluto–Charon rotates on its axis so slowly that two spacecraft would be needed to map both sides.[39]

However, for PFF to reach Pluto quickly – one of its prime goals – it would have to be more than just light; it would also have to be launched aboard one of the two most powerful launch vehicles in the mid-1990s US inventory: the Titan or the Shuttle. Titan was far preferred by most of Staehle's design team and by most of OPSWG, because it was far less expensive to integrate an uncrewed spacecraft with an uncrewed rocket, than with the crewed Shuttle, where increased safety demands ratchet up the cost.

Nonetheless, the US Air Force had a monopoly on Titan's services and wanted to charge NASA almost $500 million for *Cassini*'s launch. At that price, PFF's two rockets would cost almost a billion dollars – almost twice the total cost of two PFF spacecraft, their instruments, and flight operations. To NASA Administrator Dan Goldin this was an untenable balance. The cost of Titan became a Pluto killer, and something had to be done about it.

The Plutophiles met this problem with a double-pronged attack. At JPL, mission designer Weinstein found a series of Farquhar-esque trajectories that used successive Venus and Earth flybys to push PFF to Jupiter, and then on to Pluto. These trajectories would add three years to the trip, but PFF could be launched on them using any one of several smaller launch vehicles far less costly than Titan. Of course, the longer trip time added risk, particularly for spacecraft that carried little in the way of redundant systems.

At the same time, OPSWG's chairman, Stern, set out on an even more ambitious cost-cutting course, by internationalizing the mission. With this in mind, he set off in January of 1994 for Moscow in hopes of making inexpensive Russian launchers – perhaps, he hoped, even paid for by the Russian Space Agency – another option.

In Russia's capital, Stern gave briefings on the Pluto mission to over 50 leading space scientists and program managers at the main astronomical center of Moscow's largest university, Moscow State's Shternberg Institute, and at Russia's premier center for scientific space exploration, the Institute for Cosmic Research (whose Russian acronym, IKI, is pronounced "ee-*key*").

The pivotal moment of the trip came after the briefings, on a gray winter's afternoon in IKI director Alec Galeev's office (Figure 7.11). Stern wanted to know from Galeev ("Gal-ye-ef") whether Russia might be interested in contributing one of its powerful launch vehicles in exchange for Russian participation in the mission. Galeev response was, simply, "No," because he did not see much in NASA's Pluto mission for Russia, or Russian scientists. Stern wondered what the Russian's might consider a useful "piece of the pie," and recalled the one real aspect of the old Mariner Mark II Pluto mission concept. Stern asked Galeev whether IKI and the Russian space science community might be more interested if PFF carried a Russian-built probe designed to enter Pluto's atmosphere. Galeev sat up in his chair, smiled, and said, "*This* is an interesting idea." To Galeev, the

Fig. 7.11: Alec Galeev in about 1994, director of the Russian Institute for Cosmic Research, IKI, in Moscow. (A. Stern)

idea of a Russian Pluto entry probe offered several attractions. First, there would be the highly visible nature of the brief but daring probe mission. Second, it would give the Russian space science community a unique and valuable dataset to call their own. But most importantly to Galeev, this might be the way to move Russia's planetary exploration program into the outer solar system for the first time, "...by sending key engineers to the US to learn what NASA had mastered with the Pioneers and Voyagers."

Expresso

Within weeks, the seed that Stern and Galeev planted in Moscow took root. Stern brought the concept home to NASA and JPL. Staehle and his team were excited about the prospects for a joint mission; so was NASA's Wes Huntress, who had by then been promoted to the position of being NASA chief of all space science. Huntress briefed NASA Administrator Goldin on the concept, and found him enthusiastic as well.

Within a few more weeks, tentative letters were exchanged between Moscow and Washington suggesting that *official* exploration of a joint mission take place. Over the next few months, US and Russian teams exchanged data and developed a technical plan for the joint mission concept. The plan called for the two planned US Pluto spacecraft to be launched by either the large Russian Proton or the somewhat smaller and less expensive Russian Molniya rocket. Each of the two US PFF spacecraft was also to carry Russian probes called "Drop Zonds"[40] that would dive into Pluto's atmosphere as the tiny mother ships hurtled overhead.

While these US–Russian cooperation plans were being sketched out, Stern and JPL's Pluto mission scientist, planetary astronomer Richard Terrile, made a ten-day round-robin tour of European space centers describing the Pluto mission and investigating whether other potential partners might be interested in joining in PFF as well.

The goal of these meetings was to further internationalize the mission in order to reduce costs. Terrile and Stern were joined for part of the trip by space physicist and OPSWG member David Young. The three scientists gave briefings on Pluto mission planning and scientific objectives to German, Dutch, Swedish, French, Italian, and Spanish scientists.

By the end of 1994, the Pluto mission cabal was planning on Russian cooperation, and conducting a detailed study of a German proposal to provide a Drop Zond-like probe to study Io's atmosphere during the planned Jupiter flyby as well. Why this? Russia demanded payment for their rockets, but NASA was prohibited by law from doing so. German space scientists stepped in with a proposal to pay the Russians, in exchange for German participation and this Io probe.

PFF's main focus was moving from feasibility studies to tactical planning for what NASA officially refers to as "new start," when it solicits approval and funding to begin building a space mission. However, just as that shift in focus was beginning, NASA's Goldin mandated further reductions in the mission's size and cost. Goldin wanted to see a lighter spacecraft that would better showcase advanced technologies. Despite this being their third re-design for Goldin, Staehle and his JPL team responded by re-examining every gram of structure for savings, and by using wherever they could advanced communications, power generation, and propulsion components barely off laboratory testbeds. Staehle also enlisted the help of a University of Colorado spacecraft operations team led by Elaine Hansen. Hansen was famous for her innovative, low-cost approach to mission operations.

Fig. 7.12: An artist's impression of the Pluto Express spacecraft with Pluto and Charon in the background. (JPL)

With Hansen's help, Staehle and his team found they could use high levels of computer automation to operate Staehle's elegantly simple Pluto mission with as few as ten people at JPL, including trajectory designers, flight controllers, and the mission chief. By contrast, missions like Voyager, Galileo, and Cassini had been staffed by hundreds (and at their peak, almost a thousand) JPL flight operations personnel. By today's standards, this is a pretty standard approach but in 1994 typical deep-space mission teams involved many dozens (and sometimes even hundreds) of people – it was for its time a radical approach.

Working hard through 1995, Staehle's team blended together more than 20 advanced technologies, a new mission operations concept, and a sparing mission management concept that would set a new standard for deep-space missions. When he and his team were done, they had whittled the spacecraft mass down to 75 kilograms – without

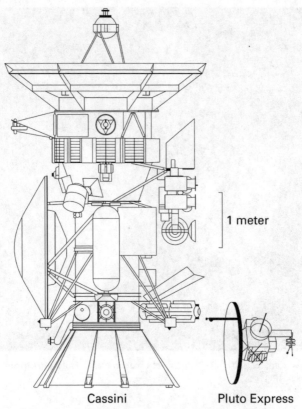

1 meter

Cassini   Pluto Express

Fig. 7.13: A size comparison between the *Cassini* spacecraft (for exploring the Saturnian system) and the proposed Pluto Express. (JPL)

changing the payload mass – and the spacecraft's power requirements to just 75 watts, which is about that of a single household light bulb. They also chose a new name to symbolize the new approach – Pluto Express (Figures 7.12 and 7.13).

Staehle's team produced detailed cost estimates and pegged the cost of developing both the Pluto Express spacecraft and their instruments, plus all of the necessary launch preparations and the flight operations system, at just under $350 million. Adjusting for inflation, that cost was less than one quarter that of Voyager. But it was a radical departure from business as usual. After all, many NASA Earth orbiter missions cost more to build than the Pluto mission

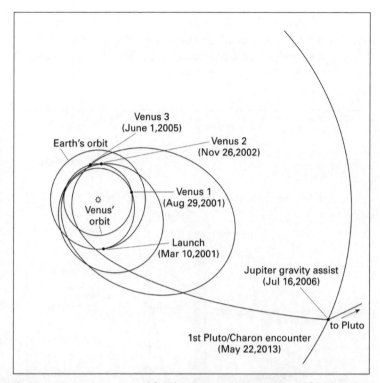

Fig. 7.14: A trajectory proposed for *Pluto Express*. This example was for a launch on March 10, 2001. It relied on three close encounters with Venus, and one with Jupiter, taking repeated advantage of the "gravity assist" technique to accelerate the craft on its way to Pluto.

would. Many experienced space project managers doubted it could be done.

But Staehle argued it would work. If NASA could muster the funding, he said, Pluto Express could be ready to launch between 2001 and 2003. Each of the two Pluto Express spacecraft would then use Farquhar/Weinstein-style swingbys of Venus and/or Earth (depending on the exact launch date) to increase its energy enough to catch a close Jupiter swingby (Figure 7.14). At Jupiter, each Pluto Express spacecraft would concentrate on the volcanic moon Io, with its German entry probe serving as the focus of the scientific investigations there.

The close pass each Pluto Express would make to Jupiter would propel it forward on a nearly beeline trajectory to Pluto. With an average speed of about 60 000 kilometers per hour, the two spacecraft would then take about 7.5 years to cross the great, yawning gulf of the outer solar system.

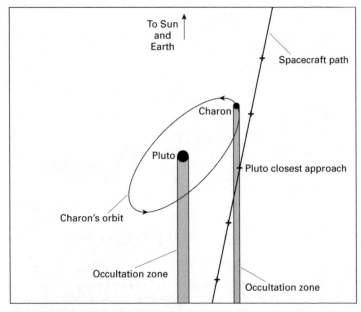

Fig. 7.15: The path *Pluto Express* would have taken (top to bottom) during its flyby encounter with Pluto and Charon. The crosses mark intervals of 20 minutes.

The first and second Pluto flybys would be about 6 months apart, so that scientists could optimize the second flyby based on what was found by the first. One spacecraft would concentrate on studying one side of Pluto; the other spacecraft would concentrate on the opposite side of the planet (Figure 7.15). Each would survey Charon and each would aim a Russian Drop Zond into Pluto's atmosphere. Each spacecraft would measure the composition and structure of Pluto's atmosphere, and map the surface geology and composition of the side of Pluto (and Charon) it would fly by.

Additionally, each spacecraft would search for small satellites in the Pluto–Charon system, refine mass and density measurements of Pluto and Charon, and actually determine the rate at which Pluto's atmosphere is escaping to space. The Russian Drop Zonds would carry a mass spectrometer deep into Pluto's atmosphere to bring a new dimension to measurements of the atmosphere's chemical composition, and make measurements to determine whether Pluto has a magnetic field.

If all went well after passing Pluto, one or both Pluto Express spacecraft would then be sent on to make flyby mapping and composition studies of one of the myriad Kuiper Belt mini-planets that orbit in the frigid wilderness beyond the ninth planet.

8

New Horizons

"The rumors of my death have been greatly exaggerated."
– Mark Twain

A Final Faltering

Sometimes something just is not meant to be. In the early 1990s
NASA studied and then put aside Pluto missions called Pluto 350,
Mariner Mark II, and Pluto Fast Flyby. Out of those ashes came
Pluto Express, which morphed into Pluto-Kuiper Express (PKE) when
the impetus to study Kuiper Belt Objects (KBOs) became almost as
important in the scientific community as the drive to study Pluto. By
the time the new millennium arrived, NASA had spent 10 years and
over $250 million on Pluto mission studies and advanced hardware
development. A cynic might conclude that NASA had appeased the
scientific community without ever intending to build and fly an actual
mission to the ninth planet.

But a more rational point of view is that the drawbacks of a mission to Pluto were simply too great for NASA to swallow. Drawbacks
you say, what drawbacks? Perhaps the most stark was (and is) the
fact that the combined time to build and fly any craft across the solar
system from terra firma to Pluto, roughly 15 years, is just too much
for a bureaucracy to invest in to achieve a singular scientific goal.
After all, any NASA bureaucrats in high enough positions to initiate a Pluto mission would certainly be retired or otherwise gainfully
engaged a decade and a half hence, making the reward something
some nameless, faceless future office holders would revel in, but not
the initiators themselves.

Moreover, bureaucracies do not like risk and a mission that would
take 10 years – or longer – just to deliver its first return is a risk
NASA has never before attempted. Most NASA spacecraft are ready
to return rewards on reaching Earth orbit. Most planetary probes
take less than three years to reach their targets. Extreme outliers like
the *Galileo* Jupiter orbiter and the *Cassini* Saturn orbiter take six or

Pluto and Charon, S. Alan Stern and Jacqueline Mitton
Copyright © 2005 WILEY-VCH Verlag GmbH & Co. KGaA, Weinheim
ISBN: 3-527-40556-9

seven years to reach their destinations. No mission in NASA's almost 50-year history has taken 10 years to deliver the goods.[41]

With these and other drawbacks, it is no wonder that concept after concept failed to reach fruition. The Plutophiles took a long time to recognize this, being lost "in the forest for the trees," if you will, throughout the early 1990s. Their initial success at obtaining mission study funding, at enthusing NASA Administrator Goldin with the romance of exploring the most distant known world, and the seemingly concrete step of being funded to develop miniaturized instruments designed specifically for their Plutonian pilgrimage conspired to give many the feeling that they were always just one step away from victory, which meant a funded project that would actually build a mission rather than just study it.

By the turning of the millennium it had been a long time – a dozen years – since the Pluto Underground had first convened in a smoky Italian diner in Baltimore, and more than six years since NASA had moved Pluto exploration to its "front burner."

By mid-2000, Pluto-Kuiper Express, though descoped (to save money) from a two-spacecraft mission to a single spacecraft, was shaping up to be a highly ambitious scientific exploration mission. It would provide a new look at the Jupiter system, a reconnaissance flyby of the Pluto–Charon system, and even a daring extended mission into the ancient realm of the Kuiper Belt.

When the Pluto Underground came together and began their quest to have NASA mount a reconnaissance mission to Pluto, they created and often used the climbing of Everest – Earth's highest peak – as a metaphor for the exploration of Pluto by spacecraft. What they did not realize when this analogy was coined, is that there are actually two Everests that any Pluto mission must scale. There is, of course, the actual journey to the Everest that is Pluto. But there is also the equally difficult journey to ascend the Everest of NASA and Washington politics. Maddeningly, by 2000, some 12 years after beginning their trek, it had taken more years to get the Pluto mission out of the Washington, DC, beltway than the mission would need to cross the whole solar system, and still there was no firm start date.

For a moment it looked as if NASA might proceed, for the space agency officially solicited instrument proposals and convened selection boards to choose the cameras, spectrometers, and other sensors to fly aboard PKE. But then, on the earthly Everest of Washington politics, Pluto-Kuiper Express, née Pluto Express, née Pluto Fast Flyby, née Mariner Mark II, née Pluto 350 faltered. With a build/no-build decision needing to be made in order to make the Jupiter launch window that would close about five years hence, and costs ballooning to over a billion dollars for a single spacecraft mission, NASA withdrew. It forcefully and publicly cancelled PKE in the autumn of 2000, walking briskly away with statements that made clear that there would be *no* successor attempt. The then NASA administrator for all of space science, Ed Weiler, declared the mission "Over, cancelled, dead."

We Are the Undead

The scientific community did not take Weiler's verdict lying down. For with the demise of PKE it was clear there was nothing to lose. More than a decade of work had been scrapped with a single pen stroke and pronouncement. It was liberating in an odd way. Previously, the scientific stakeholders had been beholden to NASA's Plutonian vicissitudes, not wishing to damage the chances of a future mission that might one day arise. But Weiler's statements made clear there was no future for Pluto missions anytime soon. It was over, he said.

The first tactic that the Pluto community adopted was to publicize broadly the cancellation of PKE and NASA's comments about the lack of any desire to find a way forward. As a result, dozens of stories appeared in the press. From *Space News* to the *New York Times*, the news reverberated around the national (and even international[42]) media echo chamber.

These stories caught the eyes of politicians (who seized on the high cost of a decade of engineering studies now laid to waste), of NASA's advisory groups, and of a young man named Ted Nichols (Figure 8.1). Nichols was only a senior in a small-town American high school in Pennsylvania but he could not believe that the USA would turn its back on the exploration of the last planet. So what did he do in the early autumn of 2000? He turned to the internet and set up a "Save the Pluto Mission" website. Aided by popular press stories about his quest, Nichols generated more than 10 000 letters to NASA in just two weeks. No one could recall the last time a space science success

Fig. 8.1: Ted A. Nichols II in 2000. He organized a popular protest against NASA's cancellation of the Pluto-Kuiper Express mission. (Courtesy Ted A. Nichols II)

had generated such a flood of public responses, *much less* a mission cancellation!

Space interest groups like the Planetary Society added their weight to the outcry at the cancellation of PKE. Scientists wrote editorial opinion pieces. Editors of major space-related publications like *Aviation Week and Space Technology*, *Astronomy Magazine*, *Sky and Telescope*, and *Space News* added to the din.

Within barely a month of Weiler's cancellation of PKE, an idea was already emerging to resurrect a Pluto–Kuiper Belt (PKB) mission. Apparently no one is certain who suggested it first. Perhaps it was even hatched within NASA. The concept was simple. If five successive JPL

Pluto mission studies had faltered, let the full aerospace community compete with concepts for the mission. NASA had by the mid-1990s turned to competition as a way to provide the government with added value and better options for many kinds of scientific space missions, and this idea was a natural extension of that trend.

By December of 2000, NASA's senior advisory panels in space science, the Solar System Exploration Subcommittee (SSES) and the Space Science and Applications Advisory Committee (SSAAC) weighed in favoring the idea. Although the idea of a competition for a Pluto mission with a single scientist (the "Principal Investigator") at its helm gained popularity, NASA pointed out that such an approach was fraught with difficulties. They cited the long nature of the mission. They cited the higher than usual price tag. They cited the need to use nuclear power aboard the spacecraft. But the scientific community, the press, the aerospace community, the Planetary Society, and Ted Nichols's band of supporters all backed it.

The week before Christmas 2000, Weiler held a press conference at NASA headquarters announcing that NASA would conduct a quick-reaction call for proposals for a PKB reconnaissance mission. Proposals would be due in March 2000 and would be cost capped at just over $500 million, less than half the cost of the $1.1 billion price tag of PKE when it was cancelled. Weiler made it clear that he was not putting himself or NASA under any obligation actually to select a mission unless it was (a) able to meet the scientific goals NASA laid out, (b) cost no more than $500 million, and (c) showed itself to be buildable and flyable with no more risk than more modest planetary missions NASA had selected this way.

Asked for a response to Weiler's press conference, former OPSWG chair and Pluto mission advocate Alan Stern simply said, "We are the undead."

A Reincarnation

Normal NASA mission competitions take place about a year after NASA announces an opportunity. With only three months between Weiler's announcement and the planned due date for PKB mission proposals, there was no time to dally. Over the holiday season, a series of six industry–academia consortia formed.

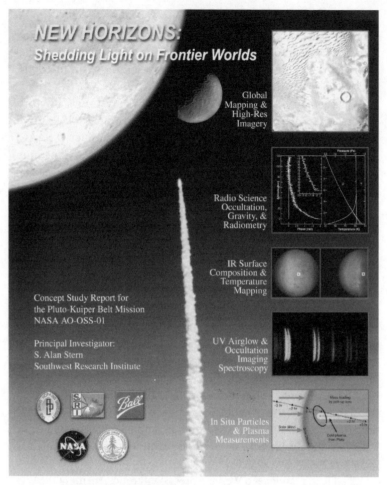

Fig. 8.2: The cover page of the New Horizons proposal to NASA. (Johns Hopkins University Applied Physics Laboratory)

Only five eventually survived the grueling pace to put together mission designs, instrument package concepts, detailed schedules and cost information, education and public outreach plans, small/disadvantaged business plans, and all of the other data NASA requested. The submitted proposals, which described all of this information, were massive – about the size of a big city's phone book (Figure 8.2).

NASA evaluated these proposals in the spring of 2001, announcing on June 6 that it had selected two of the competitors for more detailed

Fig. 8.3: The New Horizons mission logo, which shows an artist's concept of the spacecraft flying over Pluto and incorporates the logos of the major partners in the project. (Johns Hopkins Applied Physical Laboratory)

studies and a second round of proposal evaluation to come in the fall. The two winners were:

- *Pluto and Outer Solar System Explorer (POSSE)*. Dr Larry Esposito, Principal Investigator, University of Colorado, Boulder, leading a team including the following major participants: NASA's Jet Propulsion Laboratory (JPL), Pasadena, California; Lockheed Martin Astronautics, Denver; Malin Space Science Systems, Inc., San Diego; Ball Aerospace Corp., Boulder, Colorado; and University of California, Berkeley.
- *New Horizons*. Dr S. Alan Stern, Principal Investigator, Southwest Research Institute, Boulder, Colorado, leading a team including the following major participants: Johns Hopkins University Applied Physics Laboratory, Laurel, Maryland; Ball Aerospace Corp.; Stanford University, Palo Alto, California; and NASA's Goddard Space Flight Center, Greenbelt, Maryland; and JPL (Figure 8.3).

Fig. 8.4: Most of the members of the New Horizons science team (and team collaborators, indicated with asterisks) in the summer of 2004. Front row (left to right): Fran Bagenal, Mike Summers, Leslie Young, Alan Stern, Bonnie Buratti, Randy Gladstone. Middle row: Jeff Moore, Rick Binzel, Len Tyler, Andy Cheng, Will Grundy, and Hal Weaver. Back row: Ralph McNutt, Marc Buie*, Bill McKinnon, Darrel Strobel, Dennis Reuter, Rich Terrile*, and Dale Cruikshank. (Courtesy Dale Cruikshank)

With about a half a million dollars in NASA study money each, the POSSE and New Horizons teams worked all summer to provide even more detailed plans for their mission proposals. Each proposal team involved a few hundred technical experts, ranging from scientists to instrument builders to spacecraft designers to rocket designers to orbital mechanicians to spacecraft operations and tracking experts; so too, each team included accountants, managers, and educators (Figure 8.4).

Both POSSE and New Horizons planned a single spacecraft flyby mission to Pluto and the Kuiper Belt. Both teams proposed to carry cameras, spectrometers, and other instruments to survey these bodies. Both teams proposed to launch in late 2004, with a 2006 backup, traveling to Pluto via Jupiter to speed up the route (Figure 8.5). Both teams proposed to develop a new, robotic spacecraft to make the journey, but each proposed to adapt subsystems from previous spacecraft their lead institutions had developed in order to reduce the develop-

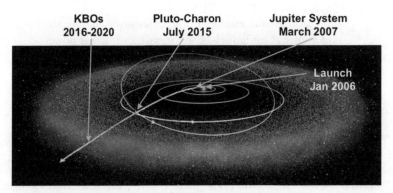

KBOs
2016-2020

Pluto-Charon
July 2015

Jupiter System
March 2007

Launch
Jan 2006

Fig. 8.5: The planned trajectory of New Horizons to Pluto and beyond for a launch in 2006. (Johns Hopkins University Applied Physics Laboratory)

ment and cost risks. Both mission proposals came in near the cost cap.

Proposals were due in late September of 2001. A month later, NASA conducted review team evaluations of the proposals, including a site visit to the spacecraft suppliers for each team, quizzing them in depth on the details of their respective proposals.

By the end of October, NASA had collected all of the information it would need to make a selection. The decision largely came down to an evaluation of which team would be most likely to be able to build and fly their mission with a lower risk of failure.

Then, on the afternoon of November 29, 2001 NASA announced its decision in a press release:

> "NASA has selected a proposal to proceed with Phase B (preliminary design studies) for a PKB mission, intended to explore the most distant planet in the solar system. The mission will also explore the Kuiper Belt beyond Pluto, a source of comets and believed to be the source of much of Earth's water and the simple chemical precursors of life. The scientific value of this mission is highly dependent on a 2006 launch that achieves a flyby of Pluto well before 2020. In order to ensure this launch date, NASA has established two conditions that must be successfully met at the conclusion of Phase B. First, the mission must pass a confirmation review that will address significant risks

such as schedule and technical milestones and regulatory approval for launch of the mission's nuclear power source. Second, funds must be available. Congress provided $30 million in fiscal 2002 to initiate PKB spacecraft and science instrument development and launch vehicle procurement; however, no funding for subsequent years is included in the administration's budget plan."

Ed Weiler said in the release, "Both proposals were outstanding, but New Horizons represented the best science at Pluto and the Kuiper Belt as well as the best plan to bring the spacecraft to the launch pad on time and within budget."

The Undead Never Blink

No sooner did New Horizons get underway in early 2002 than did the next year's budget appear in a submission to Congress, with New Horizons cancelled. What was the official reason for this? Because Pluto-Kuiper Express (yes, PKE) had grown in cost so much. No it did not make any sense, and many concluded that the Office of Management and Budget simply wanted Pluto exploration axed, despite NASA's selection of a mission costing less than PKE and Congress's funding of preliminary design studies.

The scientific community, the Planetary Society, the New Horizons mission team, and Ted Nichols went to work to save Pluto exploration once again. More editorials appeared in leading space publications supporting New Horizons and Pluto exploration. A public poll showed Pluto exploration as popular as Mars exploration. That same year, the National Research Council's Decadal Survey for Planetary Science released its ranking of mission priorities for the coming decade, 2003 to 2013, with a PKB mission at the very top of the list. Strong support in the US Congress from Senator Barbara Mikulski, then the chair of the Senate Appropriation Committee that oversees NASA's budget, was key in this. Numerous other senators and House members also supported the mission and later that year Congress sent the president a budget that fully funded the development of New Horizons. When the next NASA budget appeared, in

February 2003, New Horizons was fully funded and set on a course to launch in January 2006.

New Horizons, Indeed

New Horizons got its name in February of 2001, denoting its job of exploring a new part of the solar system. The mission concept is simple: build the smallest, lightest spacecraft that can reliably make the journey to, and first reconnaissance of, Pluto (Figure 8.6). Then place that spacecraft atop a very powerful rocket in order to minimize the trip time.

Fig. 8.6: An early full-size mock-up of the New Horizons spacecraft illustrating how compact it is. (Fran Bagenal)

The scientific requirements – the things New Horizons must do to be successful – were laid out by NASA in its 2001 request for proposals to reconnoiter Pluto–Charon and the Kuiper Belt. They are: to map the surfaces of Pluto and Charon, to map the surface compositions of these worlds, and to assay the composition and structure of Pluto's tenuous but fascinating atmosphere. Beyond these basics, the mission also intends to achieve several other objectives, such as stereo mapping to elucidate topography, thermal mapping across the surfaces of Pluto and Charon, searches for additional satellites in the Pluto–Charon system, and studies of Pluto's ionosphere. In short, the mission is designed to provide a complete first-order picture of the ninth planet, its moon, and the KBO targets it also plans to reconnoiter.

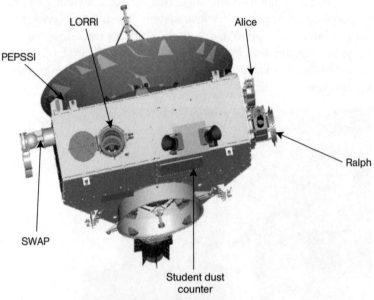

Fig. 8.7: The New Horizons spacecraft showing the location of its instruments. (Johns Hopkins University Applied Physics Laboratory)

To accomplish its scientific objectives, New Horizons will carry a suite of seven scientific instruments (Figure 8.7). The primary remote sensing devices are Ralph and Alice (named after the main characters in a 1950s television show, *The Honeymooners*). Ralph is a capable, four-color mapping camera combined with an infrared spectrometer designed to map surface compositions and temperatures. Alice is an ultraviolet spectrometer, the main purpose of which is to study the structure and composition of Pluto's upper atmosphere. Another camera, called LORRI (Long Range Reconnaissance Imager) is also included in the payload. LORRI, which is fed by a long-focal-length telescope, is less versatile than Ralph but has higher resolution. LORRI will allow New Horizons to study Pluto from a much greater distance than Ralph will. LORRI will also provide higher resolution images than Ralph during the closest portions of the flyby. Ralph, Alice, and LORRI are each sensitive enough to obtain their images and spectra even though the sunlight incident upon Pluto's surface is approximately 1000 times fainter than that received at the Earth.

New Horizons will also carry a pair of plasma sensors, called PEPSSI and SWAP, which are designed both to study material flowing off Pluto's atmosphere and to measure the escape rate of Pluto's atmosphere. Also aboard will be a radio science experiment called REX to probe Pluto's lower atmospheric density and temperature profiles, and to measure accurately the individual masses of Pluto, Charon, and KBOs. Finally, New Horizons will carry the first ever planetary science instrument built by students – the Student Dust Counter (SDC). SDC is designed to measure the density of dust particulates across the deep outer solar system and into the Kuiper Belt – something never done previously.

The total mass of the New Horizons payload is less than 30 kilograms, and the full instrument suite draws less than 30 watts. This is far more instrumentation than PKE or PFF planned to carry. Simply put, on PFF or PKE there would have been no LORRI, no SWAP, no PEPSSI, and no SDC, as only cameras, spectrometers, and a radio science package were planned.

The New Horizons spacecraft has a planned total mass of 481 kilograms, including maneuvering fuel. The spacecraft subsystems share a design heritage with other NASA planetary missions built by the Johns Hopkins Applied Physics Laboratory, such as the NEAR-Shoemaker asteroid rendezvous spacecraft that orbited 433 Eros in 2000 and the MESSENGER mission now on its way to orbit Mercury beginning in 2009.

New Horizons is designed to be able to operate at distances as far as 50 AU from the Sun, returning its data to Earth over a radio link. Almost all of the subsystems in the spacecraft have redundant capability, providing important backup provisions in the case of a systems failure during the long journey across the 5-billion-kilometer route to Pluto–Charon. The plan is to launch New Horizons aboard an Atlas V launch vehicle (Figures 8.8 and 8.9) in January of 2006, in order to use a gravity assist maneuver at Jupiter in 2007. This will shorten the travel time by over three-and-a-half years compared to the time a direct trajectory to Pluto would take.

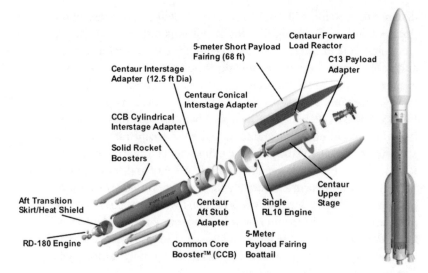

Fig. 8.8: The Atlas V launch vehicle that it is planned will launch the New Horizons spacecraft. It is about 200 feet (63 meters) tall. (Lockheed Martin Astronautics)

At Jupiter, New Horizons will conduct an intensive, six-month campaign of Jovian system observations in 2007. Closest approach will occur in March 2007 at a distance of about 30 Jovian radii (which is set by the position of the Pluto aim point at Jupiter); this is over four times closer than Cassini's Jupiter flyby in 2000–2001. New Horizons's Jupiter encounter will make possible new studies of Jupiter's atmospheric and auroral dynamics, new observations of the Galilean and irregular satellites, and valuable *in situ* exploration of the Jovian magnetosphere that will include a fortuitous 100 000 000-kilometer flight down Jupiter's long magnetotail.

After its Jupiter encounter, New Horizons will fly for almost eight more years to reach Pluto. During this time the spacecraft will hibernate most of each year, waking up only for course corrections and 60-day-long annual checkouts of its systems and instruments. The spacecraft will monitor the solar wind and energetic particle environment along parts of this journey using SWAP and PEPSSI, and using SDC it will count dust particle impacts it encounters across the outer solar system.

New Horizons is targeting a July–August arrival at Pluto–Charon. This is when Pluto lies almost directly opposite the Sun as seen from

Fig. 8.9: An Atlas V launcher leaves the same launch pad in Florida from which New Horizons will depart. (Lockheed Martin Astronautics)

the Earth, and is the optimal circumstance for minimizing solar and solar wind interference for the radio science experiment. The precise arrival day will then be chosen to place Charon behind Pluto (which occurs once every 6.4 days) so that light from Charon will softly illuminate Pluto's night side for imaging.[43]

New Horizons's Pluto–Charon encounter is planned to begin about six months prior to closest approach. At this time, Pluto will be just a little farther from the spacecraft than the Sun is from the Earth – 1 AU. At first, Pluto will be little more than a bright point of light ahead, but New Horizons will train Ralph and LORRI on the approaching binary world and then begin taking spectra and test images. At the same time, SWAP and PEPSSI will measure the charged particle environment surrounding Pluto. SDC will measure the number and masses of dust particles in the region as well by counting the frequency and energetics of dust impacts on the spacecraft.

Then, for a period of 75 days on either side of closest approach, the resolution of images of Pluto and Charon made by LORRI will exceed the best that the Hubble Space Telescope ever achieved. Early images and other data will allow mission planners to optimize the close approach sequence and to obtain a long series of disk-resolved images to study time-variable phenomena, such as the movement of frosts across the surface and meteorological phenomena, before closest approach. During this time LORRI will also search for possible tiny moons of Pluto that could still remain undiscovered from Earth; if one or more are found, its orbit will be determined and mission planners will train the full battery of imagers and spectrometers that New Horizons carries on the newfound worldlet(s). Beginning a few days before closest approach, SWAP and PEPSSI will be used to measure the rate and composition of gas that is escaping from Pluto's atmosphere.

Closest approach is currently planned to be at the half-way point between Pluto and Charon but could be as close as only 3000 kilometers above Pluto. In the hours before closest approach, New Horizons will obtain maps of both Pluto and Charon with kilometer-scale resolution, color maps with somewhat lower resolution, and spectral maps that will determine Pluto's surface composition at 50 000 to 1 000 000 locations. The same datasets will be collected on Charon. At closest approach itself, images of Pluto with resolution as high as 25 meters per pixel may be achieved, depending on the exact flyby distance selected.

The encounter is being designed to probe Pluto's atmosphere with two special occultation experiments. In the first, radio signals detected by NASA's Deep Space Network on Earth will be recorded as New Horizons flies behind the planet. The radio science experiment, REX, will measure the amount of refractive bending of these signals as a function of altitude above Pluto to determine the temperature–pressure structure of the lower atmosphere; REX will also be able to determine, or set a good limit on, Pluto's ionospheric density. During roughly this same time period, the Alice ultraviolet spectrometer will observe a sunset and sunrise from behind Pluto in order to obtain a profile of how the composition, temperature, and density of Pluto's atmosphere vary with altitude and location.

Fig. 8.10: New Horizons under construction in early 2005. This image shows the 2.1-meter main antenna after its installation on the spacecraft. (NASA/Johns Hopkins University Applied Physics Laboratory/Southwest Research Institute)

As New Horizons exits the Pluto–Charon system it will radio about 10 gigabytes of data back to Earth, containing all of the spectra, maps, and other measurements it made. About three weeks after passing Pluto–Charon, New Horizons will fire its engines to put itself on a course for its first KBO flyby. Studies indicate that New Horizons may be able to reach one or even two KBOs in the years following its encounter with Pluto–Charon, depending on the amount of fuel in its tanks and the health of the spacecraft.

Light My Fire

As we write these words New Horizons is being built (Figures 8.10 and 8.11). Its scientific instruments are almost complete. NASA is applying for launch permission for the January 2006 Jupiter launch window and a 2007 backup trajectory as well. The New Horizons team has invited both Clyde Tombaugh's widow, Patsy, and Venetia Burney, who as a girl named Pluto, to the launch, along with virtually every astronomer who has studied the Pluto–Charon system.

Fig. 8.11: The LORRI instrument being installed on New Horizons in the clean room at Johns Hopkins University Applied Physics Laboratory. (NASA/Johns Hopkins University Applied Physics Laboratory/Southwest Research Institute)

By the time of launch it will have taken 17 years or more to see the dream of a mission to the Pluto–Charon system come to fruition. That is a long time – far more time than it will take New Horizons to cross the solar system to its farthest frontier – but good ideas sometimes take time to mature. If all goes well as New Horizons is built and tested, there will soon come a day when its builders will see it lofted heavenward to be the first mission to the so-called "last planet."

9
Where No One Has Gone Before

"Ad Astra per Aspera (to the stars through difficulties)."
– Kansas State Motto

For over four billion years Pluto and Charon orbited the Sun in cold and quiet anonymity. But in the last fraction of the last percent of that span came something new. That something happened on the third planet, and it was the rise of *our* species – the first in Sol's solar system to cast its gaze and intellect to the heavens.

What drives this thing in just one species among the myriad life forms that Earth has produced in four eons? Why do we, the people of Earth, need to explore? We do not know the basis behind this drive, but we do know it makes us feel alive!

And when we, *Homo sapiens*, looked beyond our own planet, we found a universe matching our greatest hopes and aspirations. Eventually we found a world that, at first certainly, seemed only a footnote to our universe and our planetary system – Pluto.

Somehow, however, the smallest of the known planets took on a special attraction, apparently out of all proportion to its comparatively gargantuan companions. Perhaps it was just the mystery of this little world so far away, or perhaps it was Pluto's diminutive size and underdog position among the planets that made it in many ways what the Russians call *nasha dyevoshka*, our dearest little girl.

What a wonderful little world Nature had concocted out there from a few ices and minerals, a little classical low-temperature physics, and a time span a thousand times our own.

Still, after over 75 years of study, so many of Pluto's secrets remain, guarded by the protection that its distance and diminutive size so uniquely impose. Yes, it is a challenge, a long road, an uphill battle! But we astronomers and space scientists (the heaven-looking ambassadors of our inquisitive species) continue to work to tear the covers from those secrets, and expose them to the light.

Pluto and Charon, S. Alan Stern and Jacqueline Mitton
Copyright © 2005 WILEY-VCH Verlag GmbH & Co. KGaA, Weinheim
ISBN: 3-527-40556-9

Then I Saw Her Face

As we probe the planets and the stars and the galaxies with all our wit, and the newest tools of our technology, we often succeed – though the road of exploration is rarely a straight one. And sometimes we even fail. But the best times are when a new, hard-won fact reveals a dozen even newer questions beneath it.

Just such a day came on March 8, 1996, for this was the day that world saw the true face of Pluto for the first time.

It was the best of our long eyes that did it – the celebrated Hubble Space Telescope. Thanks to its perch, high above Earth's shimmering atmosphere, the looking glass of this fantastic telescope sharpened the fuzzy little spot that had been Pluto in our telescopes – into a disk with a curiously mottled surface.

Hubble had managed something no other telescope could. After all, Pluto's diameter is just 0.1 arcseconds (1/36 000 of a degree): imaging it is as difficult as counting the spots on a soccer ball 600 kilometers (375 miles) away!

The Hubble images (Figures 9.1 and 9.2) are crude, and faint, and fuzzy, and tantalizing. They reveal that Pluto is indeed an amazingly complex object, with more intricate large-scale surface patterns than any other planet, except Earth and Mars. Pluto's surface displays a dozen distinctive provinces, some of which are more than 1000 kilometers (600 miles) across. These include a "ragged" northern polar cap, clusters of bright and dark spots in Pluto's equatorial regions, and a bright linear marking that might just be a long ejecta ray from an unusually large crater. The Hubble images and the map produced from them proved that Pluto's surface has the most complex network of features seen at this resolution anywhere in the outer solar system.

It was a wonderful achievement. But it was also a melancholy one, for those images are the best that any existing or planned telescope can do from our perch down among the warm inner planets near the Sun, 5 billion kilometers from Pluto.

And so most of Pluto's long-held secrets remain secrets – guarded by the gulf of distance between us and that ancient, icy little world. Hubble's view delivered to us a glimpse of the pinkish little ice-ball ornament, tantalizingly close in a way, but still too maddeningly far. We want to see more, but we cannot, for there is no better telescope to use. To see the solar system's "ninth sister" as it really is, we must

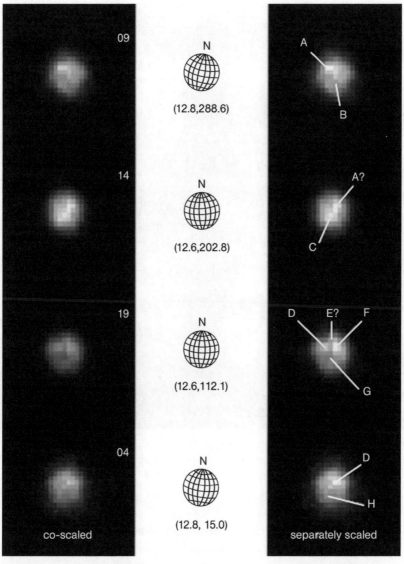

Fig. 9.1: Images of Pluto obtained with the Hubble Space Telescope in 1994, and first released after detailed image processing in March of 1996. Pluto's north pole is at the top in all images. Rotation advances from top to bottom by approximately a quarter of a turn between frames. The brightness of the images in the left column is scaled so as to preserve the relative intensity between frames. The images in the right column are scaled individually to show the maximum detail. The letters A through H label prominent features in the images. Note that some features are detected in more than one image as Pluto rotates (e.g., D). The center column shows a wire-frame grid scaled to the apparent size and geometry of Pluto for each image. Below each are given the latitude and east longitude for the sub-Earth point at the time of mid-exposure. (S. A. Stern and M. W. Buie)

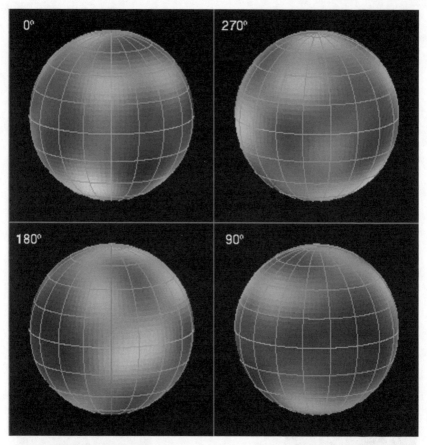

Fig. 9.2: The map of Pluto inferred from 1994 Hubble Space Telescope images projected onto four globes differing by a quarter of a turn. (S. A. Stern and M. W. Buie)

go to it. Amazingly, our species has developed the will and the way to do just that.

So guard your secrets while you can, Pluto. We are coming to wrest them from you for all humankind.

Appendix: A Chronology of Major Events in the Exploration of Pluto

1905 Percival Lowell undertakes his search for Planet X beyond Neptune. He will not succeed.

1916 Percival Lowell dies.

1929 Under the direction of Vesto Slipher, Lowell Observatory hires Clyde Tombaugh to start a new, systematic search for Lowell's trans-Neptunian planet.

1930 Pluto discovered by Clyde Tombaugh on February 18. The discovery was announced publicly on March 12–13.

1930 Pluto's orbit determined.

1955 Pluto's 6.4-day rotation period determined (Walker and Hardie).

1965 Pluto's 3:2 orbital resonance with Neptune discovered (Hubbard and Cohen).

1970 First crude spectrum of Pluto; reddish surface color found (Fix et al.).

1976 Methane ice detected on Pluto's surface (Cruikshank, Plicher, and Morrison).

1978 Charon discovered, and from that the total mass and average density of Pluto and Charon were determined (Christy and Harrington).

1978 Pluto–Charon mutual events predicted (Andersson).

1980 Stellar occultation reveals Pluto's radius is just over 600 kilometers (Walker).

1983 IRAS satellite makes infrared measurements of Pluto's surface temperature.

1985 Onset of mutual occultation events between Pluto and Charon (first unambiguous detections by Binzel and Tholen).

1986 First determination of reliable radii for Pluto and Charon (Tholen and Buie).

Pluto and Charon, S. Alan Stern and Jacqueline Mitton
Copyright © 2005 WILEY-VCH Verlag GmbH & Co. KGaA, Weinheim
ISBN: 3-527-40556-9

1986 First radio-thermal measurements of Pluto's temperature (Altenhoff et al.).

1987 Water ice discovered on Charon (Buie et al. and Marcialis et al.).

1988 Stellar occultation reveals Pluto's atmosphere (Elliot et al.).

1988 Formation of informal group interested in Pluto mission.

1989 Pluto at perihelion. Mutual events produce evidence for polar caps (Binzel).

1990 First NASA-sponsored Pluto mission study (Pluto 350) completed.

1991 Published arguments that ice dwarfs like Pluto were common in the early solar system (Stern).

1991 Pluto 350 mission concept replaced by Mariner Mark II mission concept.

1991 Outer Planets Science Working Group (OPSWG) formed.

1992 Discovery of nitrogen and carbon monoxide ices on Pluto (Owen et al.).

1992 First of many Kuiper Belt Objects discovered (Jewitt and Luu).

1992 Mariner Mark II replaced by Pluto Express mission concept.

1993 Evidence for large thermal differences across Pluto's surface (Stern, Weintraub, and Festou; also Jewitt).

1993 Pluto Fast Flyby enters phase A development.

1994 Methane detected in Pluto's atmosphere (Young et al.).

1995 Pluto Fast Flyby cancelled and replaced by Pluto Express.

1995 Charon's nonzero lightcurve amplitude discovered (Buie and Tholen).

1996 Hubble Space Telescope detects numerous surface features on Pluto (Stern, Buie, and Trafton).

1997 Clyde Tombaugh dies.

2000 Pluto Express cancelled.

2000 Evidence for ammoniated compounds on Charon's surface (Buie and Grundy).

2001 Discovery of the first Kuiper Belt binary (Villet et al.).

2001 NASA conducts a competitive Pluto–Kuiper Belt mission selection. *New Horizons* selected.

2002 Stellar occultations reveal Pluto's atmosphere has dramatically changed since 1988 (Elliot et al.).

2003 New Horizons approved for build.

Notes

1 In those days, astronomical photographs were made using glass plates coated with light-sensitive chemical emulsions.

2 The opposition point gets its name from the fact that it is the point opposite the direction of the Sun.

3 More than 65 years later, Tombaugh loved recalling how he discovered a planet, "during the daytime!"

4 March 13 Greenwich Mean Time.

5 All of the planets closer than Uranus could be seen by the naked eye and had been known since antiquity.

6 Notably, Venetia Burney was the great niece of Henry Madan, who had named Mars's moon Phobos in 1878. Venetia Burney grew up to become a teacher and on marriage became Mrs Venetia Phair.

7 In 1919 the French astronomer P. Reynaud had suggested "Pluto" as the natural mythical name for Lowell's putative Planet X, but this obscure suggestion was not remembered until years later.

8 By days we mean 24-hour Earth days; each planet's own "day" – the time it takes to rotate on its axis – has a different length.

9 Had Pluto been discovered then, and Pickering and young Humason been the heroes … But this was not the history that unfolded.

10 "Measuring plates" is shorthand astronomers use when they mean to determine the precise positions of objects on them.

11 Although the name Charon was widely used, it did not become officially recognized by the IAU until 1985 when the onset of mutual eclipses removed the last infinitesimal shred of doubt that Charon was indeed a satellite.

12 Much as Cruikshank, Pilcher, and Morrison had inferred on the basis of a bright, methane coating, but this estimate was a far more direct result, which did not depend on the assumed characteristics of Plutonian methane frost.

13 About a decade before they were available for home use in cameras and video equipment.

14 Years later, after many events had been observed, it became clear that Tedesco and Buratti's supposed signature very likely occurred outside the time that the actual event was observable. Despite the fact that most mutual event experts now believe the Tedesco–Buratti January 16 report was a false alarm, Tedesco and Buratti's contribution to the first detection of mutual events remains an important one.

15 Recall that, except under outstanding observing conditions, even the best telescopes could not cleanly split Pluto from Charon, a full arcsecond apart.

16 Had astronomers known the separate masses of Pluto and Charon, then their individual densities could have been calculated, but determining the individual masses of the pair was beyond capabilities in the 1980s. In fact, the individual masses still are not well enough known to tell definitively if Pluto and Charon have different densities.

17 Almost 15 years later, ammonia and ammonia hydrates ices were also identified on Charon's surface by multiple teams, using much more sensitive, higher resolution spectrometers than were available in the late 1980s.

18 The Kelvin temperature scale used by scientists starts at absolute zero, which is equivalent to −273 ˚C or −463 ˚F. Using the Kelvin scale avoids negative numbers for low temperatures, and provides an easy reference point to the true zero point of nature. "Degrees" on this scale are called "kelvins," abbreviated to K. A kelvin is the same as a degree Celsius, or 1.8 degrees Fahrenheit.

19 In fact, at one time NASA had planned to direct *Voyager 1* to Pluto for a 1986 flyby. But this opportunity was lost when scientists instead directed *Voyager 1* to make a close flyby of Saturn's giant moon Titan, causing the spacecraft to be slung out of the solar system on a path directed far away from Pluto.

20 S. A. Stern, L. M. Trafton, G. R. Gladstone 1988, Why is Pluto bright? Implications of the albedo and lightcurve behavior of Pluto, *Icarus* 74.

21 Which was retired in 1995, to be replaced by an event larger airborne observatory called SOFIA, in 2005. SOFIA boasts a 2.5-meter (about 100-inch) diameter telescope.

22 More familiar, lower-altitude examples of haze layers on Earth include fogs and the pollution layers over large cities.

23 Recall that the previously available, lower-resolution spectra of Pluto's methane bands could not distinguish the gas from the ice.

24 ISO was the European Space Agency's Infrared Space Observatory satellite, which operated from 1995 through 1998.

25 The KAO did not fly to these events because it was by then retired from service.

26 Astronomers use the distance from the Earth to the Sun as a convenient distance scale. This distance (149.6 million kilometers, or 93.5 million miles) is called an astronomical unit, or AU.

27 By heavy elements, astronomers mean everything but hydrogen and helium.

28 This is due to the fact that the gas molecules exert a gas pressure, which "supports" the gas, making it feel as if it is farther from the Sun than it actually is. As a result, the gas orbits a bit more slowly than the solid bodies in the disk. Hence, the solid bodies feel a headwind as they orbit through the gas.

29 The reason Jupiter grew faster and reached the hydrodynamic gas accretion stage is likely due to the fact that it is just beyond the so-called "line" where ice can condense, thereby providing an additional source of solid material to speed the growth of proto-Jupiter.

30 Of course, we are using terms that suggest the two objects were consciously hunting for one another, when in fact they were simply inanimate bodies rattling around in an imaginary box defined by their orbits.

31 This range of uncertainty is enormous, primarily because the range of possible encounter speeds was left to be just about anything consistent with objects orbiting the Sun. But if you think like an astronomer, you realize it does not matter. Regardless of the encounter speed, the time needed to have a high probability of a collision between Pluto and a single proto-Charon is far, far longer than the age of the solar system.

32 Described in a 1991 research paper by Alan Stern.

33 Known since 2004 as the Max Planck Institute for Solar System Research.

34 As did the Voyager and Pioneer spacecraft that flew by the giant planets.

35 The fact that the planets themselves remain is due in part to their wide spacings and in part to their large relative masses.

36 The IAU's nomenclature code is this: year of discovery, followed by a letter for the two-week period in the calendar when the object was discovered, followed by a letter (and if necessary a number) to indicate the order of discovery in that two-week period. Thus, 1992QB$_1$ was the 28th new object found in period 17 (i.e., Q) of 1992. (The 27 others were more mundane discoveries in the asteroid belt.)

37 Making the total transit time some 13–15 years.
38 Two more kilograms were allotted to a radio experiment that would probe Pluto's lower atmosphere.
39 If just one spacecraft were sent out, then Pluto's 6.4-day rotation period would allow it to visit only one side of Pluto close up. The other side would last be seen 3.2 days – half a rotation – before closest approach. At that point, millions of kilometers out, far fewer details would be seen.
40 Zond is the Russian term for a space probe.
41 Not even Voyager, for these craft were launched to study Jupiter and Saturn, which they reached within 4.5 years of launch. When *Voyager 2* went on to Uranus and Neptune over the next 8 years it was an extension of its mission with little risk of embarrassment, for the main mission had already been accomplished.
42 With stories in numerous European publications, including *The Economist*, the *Daily Telegraph*, the *Daily Mail*, and *Der Stern*.
43 A full Charon is almost twice as bright as the full Moon of the Earth. Although Charon is 30 times as far from the Sun as Earth's Moon, Charon is 23 times closer to Pluto than the Moon is to Earth, and its icy surface is about five times more reflective than that of our own, rocky Moon.

Index

Pluto and Charon, S. Alan Stern and Jacqueline Mitton
Copyright © 2005 WILEY-VCH Verlag GmbH & Co. KGaA, Weinheim
ISBN: 3-527-40556-9